COLUMBIA COLLEGE
W9-DFL-264
9143 9920

# CLIMATE CHANGE

ENTERED APR 1 4 2009

*Also available in the* Small Guides to Big Issues

*Women's Rights*
Geraldine Terry

*Cities*
Jeremy Seabrook

Columbia College Library
600 South Michigan
Chicago, IL 60605

**Small Guides To Big Issues**

Melanie Jarman

# CLIMATE CHANGE

DISCARD

Fernwood Publishing
Halifax and Winnipeg

Pluto Press
London • Ann Arbor, MI

palgrave
macmillan

Oxfam

JACANA

First published 2007 by Pluto Press, 345 Archway Road, London N6 5AA, UK
and 839 Greene Street, Ann Arbor, MI 48106, USA. www.plutobooks.com
Published in Australia by Palgrave Macmillan Publishers Australia, Claremont
Street, South Yarra 3141, Australia. Associated companies and representatives
throughout the world. Visit our website at www.macmillan.com.au
Published in Canada by Fernwood Publishing, Site 2A, Box 5, 32 Oceanvista Lane,
Black Point, Nova Scotia, B0J 1B0 and 324 Clare Avenue, Winnipeg, Manitoba,
R3L 1S3. www.fernwoodpublishing.ca
Published in South Africa by Jacana Media (Pty) Ltd, 10 Orange Street,
Sunnyside, Auckland Park 2092, South Africa, tel +2711 628 3200.
See a complete list of Jacana titles at www.jacana.co.za

Published in association with Oxfam GB

The views expressed are those of the author, and not necessarily those of the
publishing organisations.

Copyright © Melanie Jarman 2007

The right of Melanie Jarman to be identified as the author of this work has been
asserted by her in accordance with the Copyright, Designs and Patents Act 1988.

ISBN     978 0 7453 2581 1 (Pluto hardback)
ISBN     978 0 7453 2580 4 (Pluto paperback)
ISBN     978 1 4202 5604 8 (Palgrave Macmillan)
ISBN     978 1 55266 252 6 (Fernwood)
ISBN     978 1 77009 380 5 (Jacana)

British Library Cataloguing in Publication Data. A catalogue record for this book
is available from the British Library

Library of Congress Cataloging in Publication Data applied for

National Library of Australia cataloguing in publication data applied for

Library and Archives Canada Cataloguing in Publication
Jarman, Melanie
     Climate change / Melanie Jarman.
(Small guides to big issues)
Includes bibliographical references.
ISBN 978-1-55266-252-6
     1. Climatic changes. 2. Climatic changes--Government policy. 3. Climatic
changes--Political aspects. I. Title. II. Series.
QC981.8.C5J37 2007          363.738'74          C2007-903297-4

10 9 8 7 6 5 4 3 2 1

Designed and produced for Pluto Press by Curran Publishing Services, Norwich

Printed and bound in the European Union by
Antony Rowe Ltd, Chippenham and Eastbourne, England

# Contents

# Series Preface

## Facts, opinions, and ideas in the fight to end poverty

**Small Guides to Big Issues** is a series of accessible introductions to key current global challenges.

Books in this series raise bold questions about the global economic and political system, and about how it works. They set out what needs to happen in order to end poverty and injustice. They are designed for campaigners and activists, for students and researchers, in fact for anyone interested in looking behind the headlines.

Each book is informed by personal knowledge and passion and is written in an accessible and thought-provoking style. Each book provides a critical survey of its subject and a challenging look at current trends and debates. Authors explain the global institutions and processes involved, and tackle key issues of poverty reduction, human rights and sustainable development.

The books contain case studies, analysis, and testimony from activists and development practitioners, many drawing on Oxfam's experience of working with partner organisations in more than 70 countries. Oxfam is supporting this series by asking writers with a personal on-the-ground knowledge of the issues to share their view of the key debates in each subject.

## Oxfam GB

Oxfam GB, founded in 1942, is a development, humanitarian and campaigning agency working with others worldwide to find lasting solutions to poverty and suffering. Oxfam GB is a member of Oxfam International, a confederation of 13 organisations around the world working for an end to injustice and poverty. Oxfam is committed to making poverty-focused information and analysis of global issues more widely available, and is working in partnership with Pluto Press on this series in order to contribute to current debates. For clarity, the books in this series do not differentiate between Oxfam International or Oxfam GB, or regional offices or affiliate organisations – choosing rather to use 'Oxfam' to cover any or all of these institutions.

## Pluto Press

Pluto Press is an independent, progressive, London-based publisher specialising in books on politics and across the social sciences that offer a radical counterpoint to the mainstream. Recent titles address some of the most pressing and contentious issues today, including the economic and ecological impact of globalisation, global insecurity and terrorism, and international human rights.

# List of Boxes and Tables

## Boxes

## Tables

# List of Abbreviations

| | |
|---|---|
| ASPO | Association for the Study of Peak Oil |
| BTC | Baku-Tblisi-Ceyhan pipeline |
| CCS | carbon dioxide capture and storage |
| CDM | Clean Development Mechanism |
| $CO_2$ | carbon dioxide |
| $CO_2e$ | carbon dioxide equivalence |
| Coica | co-ordinating body of indigenous organisations of the Amazon basin |
| COP | Conference of the Parties to the United Nations Framework Convention on Climate Change |
| DMS | dimethyl sulphide |
| DOEs | designated operational entities (authorised to verify CDM claims) |
| EB | Executive Board of the CDM |
| EIR | Extractive Industries Review (of the World Bank) |
| EPICA | European Project for Ice Coring in Antarctica |
| EREC | European Renewable Energy Council |
| ETS | Emissions Trading Scheme |
| GDRs | Greenhouse Development Rights |
| HFC-23 | hydroflorocarbon |
| HIPC | Heavily Indebted Poor Countries |
| IBRD | International Bank for Reconstruction and Development (of the World Bank) |
| ICSID | International Centre for Settlement of Investment Disputes (of the World Bank Group) |

| | |
|---|---|
| IDA | International Development Agency (of the World Bank) |
| IEA | International Energy Agency |
| IFC | International Finance Corporation (of the World Bank Group) |
| IGCC | integrated gasification combined-cycle |
| IPCC | Intergovernmental Panel on Climate Change |
| IPPR | Institute for Public Policy Research |
| IQ | Inuit Qaujimajatuqangit (traditional Inuit knowledge and behaviour codes) |
| LDC | least developed countries |
| MDG | Millennium Development Goals |
| MIGA | Multilateral Investment Guarantee Agency (of the World Bank Group) |
| NAPA | national adaptation plan of action |
| NCAR | National Center for Atmospheric Research |
| PCF | Prototype Carbon Fund |
| ppm/ppmv | parts per million by volume |
| PRSP | poverty reduction strategy paper |
| SCCF | Special Climate Change Fund |
| SEEN | Sustainable Energy & Economy Network |
| UNFCCC | UN Framework Convention on Climate Change |

# Introduction

No wonder these birds are a protected species. If we let all
the wonders go we'll become crazy with our sense of loss.
                        Rose Tremain, *The Swimming Pool Season*[1]

## Two stories

On cold nights in Cape Town, South Africa, friends and
neighbours can be found gathered at the house of a man called
Muzelli.[2] While Muzelli may well be the perfect host, his
company is not the only attraction for those who have come
together under his roof: the roof itself is a large part of the
draw. Muzelli's ceiling has recently been insulated, keeping his
house much warmer than others in his neighbourhood. And
the ceiling is not the only change on this low-income housing
estate. Some houses have new solar water heaters. Others have
had low-energy light bulbs fitted for the first time. These new
developments are intended to reduce the greenhouse gas emissions from Muzelli's neighbourhood. But they are not the
result of DIY by individual homeowners, nor are they the
result of extra money from a housing corporation. No, the
drive behind these changes comes from thousands of miles
away. Funding for the low-emission projects in this Cape
Town neighbourhood was agreed by politicians of the G8
meeting in Scotland in 2005, to compensate for greenhouse
gas emissions from the world's richest countries.

In another part of southern Africa, greenhouse gas emissions from the world's richest countries have also had a direct impact on someone's life. This time, however, the effect is far more destructive.

A woman called Joanna living in Nwadjahane in Mozambique has had little chance of entertaining her neighbours as Muzelli did. Joanna lost the roof over her head when her house was lost in storms and flooding in the year 2000.[3] She rebuilt the house, but in 2004 it was badly damaged again. Undaunted, she rebuilt it again and got to work planting crops. But the terrible weather affected her work as well as her home. A drought killed her seedlings. Joanna re-planted, but the drought continued well into what should have been the rainy season. In response, she planted a drought-resilient crop the following year. Unfortunately the water situation continued to be disastrously unpredictable: this time, heavy rains brought flooding and destroyed the harvest.

While Muzelli and Joanna don't live so very far from each other, their experiences are worlds apart. What brings them together is the story that shapes our world today – the story of climate change. This story determines the state of life support systems for everything on Earth. For some species it determines whether or not they live at all. The story of climate change impacts on the availability of food and water systems. It shapes the nature of homes and livelihoods and community relations. It affects the life of every person on the planet.

## Warming trends

Back in 2005 – the year in which the G8 funded Muzelli's insulated ceiling – the story of climate change seemed to be reaching a plot climax. Temperature records were broken as the year became the hottest to date. Previous thermometer-busting years had been helped into the record books by the presence of an El Niño, a weather event where changes in ocean currents and atmospheric circulation bring altered weather patterns, including raised temperatures. The lack of this weather phenomenon made 2005's record-breaking heat even more surprising.

Perhaps not so surprising was the fact that this year of the highest temperature ever also saw the least arctic ice ever. Polar bears were found stranded at sea, exhausted and far from land, as Arctic ice cover was reduced to the smallest area that had ever been recorded.

Meanwhile, in a different part of the globe, residents of New Orleans were stranded in sports halls and on an intercity highway as their homes disappeared under floodwater during a record-breaking Atlantic hurricane season. Lester Brown, President of the US-based Earth Policy Institute, described the people displaced by Hurricane Katrina as 'climate refugees', whose ranks will grow unless the world cuts emissions of greenhouse gases.[4] Brown said: 'What we're looking at is the potential not of displacing thousands of people, but possibly millions of people as the result of rising seas and more destructive storms in the years and decades ahead if we don't move quickly to reduce carbon dioxide emissions.'[5] According to one estimate, a combination of rising sea levels, erosion and agricultural damage due to climate change could make 150 million people into environmental refugees by 2050.[6] The suggestion has been made that UN agencies draw up an internationally accepted definition of 'environmental refugee', since current treaties recognise only political refugees as eligible for aid from the UN refugee agency.

That 2005 Atlantic hurricane season broke records for the most hurricanes, the most category-five storms, the most storms hitting the United States and the most expensive hurricane damage. Those in charge of naming the storms had to work overtime – the sheer number of hurricanes exhausted the official naming system as the 2005 hurricane season also produced the highest number of named storms. Ever. It seemed that 2005 was a year in which records were broken at a rate that was, well, record breaking.

Against this background, politicians from the world's most industrialised countries, meeting as the G8, placed climate change right at the top of the agenda of their annual summit. But their meeting came up with a plan far short of what was required. Token payments, such as those for a ceiling for Muzelli, are nothing compared to the actions and funding needed to stabilise the climate. Without a stable climate, meaningful development – better housing and food security for people like Joanna – doesn't stand a chance.

## For real

Although the G8 summit back in 2005 may have been characterised by the politicians having their heads stuck in the sand, their scientific counterparts had a far more realistic perspective on the matter.

In the run-up to the G8 meeting the national science academies of the G8 countries, plus the national science academies of the five largest developing countries – Brazil, China, India, Mexico and South Africa – released a joint statement outlining grave concern over climate change. Even the US National Academy of Sciences signed up, bucking that country's trend of denying the severity of climate change. The statement got right down to business. It began with the simple phrase: 'Climate change is real.'[7] 'Warming', it said, 'has already led to changes in the Earth's climate.'[8]

The G8 statement echoed a conclusion drawn by the body that assesses climate change information for the United Nations, the Intergovernmental Panel on Climate Change (IPCC). The G8 statement identified that the root of the problem is us.

'Human activities', it said, 'are now causing atmospheric concentrations of greenhouse gases – including carbon dioxide, methane, tropospheric ozone, and nitrous oxide – to rise well above pre-industrial levels.'[9]

It's worth noting that the existence of these gases in themselves is not a problem. They are known as greenhouse gases because, like a greenhouse, they trap heat that the Earth radiates as it is warmed by the sun. This heat keeps our planet warm enough to support an abundance of biodiversity. Without greenhouse gases average temperatures would be about 30°C lower and the Earth would resemble the frozen plains of Mars. Yet when these greenhouse gases get too highly concentrated the temperature goes too far in the opposite direction, overheating the planet and upsetting conditions for life as we know it. The largest global contributor to increasing temperatures is carbon dioxide ($CO_2$) emissions. And, while land-use changes such as agriculture and deforestation account for between 15 and 30 per cent of annual global carbon dioxide emissions,[10] the largest source of the gas is our use of fossil fuels.

Greenhouse gases are not the only influence on Earth's temperature. Lower-level air pollutants such as sulphate aerosols have a cooling effect, as they reflect sunlight back into space (the increase in these pollutants from 1940 to 1970 explains why global temperature remained relatively stable during those decades, despite the industrialisation that followed the Second World War[11]). However, since these pollutants only stay around for a number of months, and carbon dioxide stays in the atmosphere for approximately 100 years, the warming from greenhouse gases eventually overwhelms the cooling from low-level pollutants.

The sun also influences Earth's climate but, again, its impact has been swamped by the build-up of greenhouse gases. Buried in the data for the IPCC's 2007 report on the physical basis of climate change are figures showing that human-made warming is at least five times greater than that due to changes in solar output. Unfortunately, all mention of this was cut from the summary text at the insistence of Chinese and Saudi Arabian delegates.[12]

While influences such as air pollutants, the sun and natural greenhouse gas emissions make the climate an unpredictable beast, the emissions caused by human activity have now pushed it far beyond any twists and turns that would occur naturally. Carbon dioxide emissions in particular have reached an unprecedented level. The science academies' statement described them as: 'higher than any previous levels that can be reliably measured'.[13] A 2001 IPCC assessment said it was likely that carbon dioxide concentrations had not been exceeded during the past 20 million years.[14]

## Rates of change

High on the hilltop of a former volcano in Hawaii sits an observatory that, since the 1950s, has monitored changes in concentrations of greenhouse gases in the atmosphere. Recent years have seen this observatory record astounding figures, and 2005 saw a record high of globally averaged concentrations of carbon dioxide.[15] However, these figures haven't just reached an unusually high point – they are also increasing at an unusually high rate. When concentration measurements were first taken in the 1950s, the annual increase was less than one part per million by volume (which abbreviates to ppmv or ppm). In 2002 and 2003, and again between 2004 and 2005, the annual increase in carbon dioxide concentrations was approximately two ppm. This figure may sound small, but relatively speaking this rate of increase is massive. It is unprecedented during at least the past 20,000 years.[16]

Commenting on figures for 2004, Charles Keeling, who had measured greenhouse gas concentrations over the previous half century, said that:

The rise in the annual rate ... is a real phenomenon. It is possible that this is merely a reflection of natural events ... but it is also possible that it is the beginning of a natural process unprecedented in the record.[17]

## Box I.1: The Keeling curve

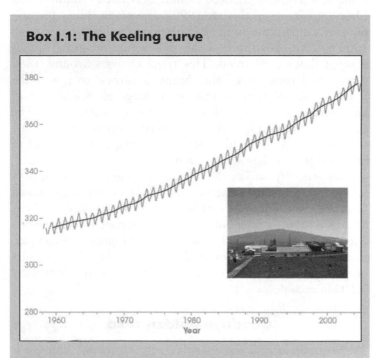

Known as the 'Keeling curve', this graph shows increasing carbon dioxide concentrations since 1958. The measurements were taken at the Mauna Loa observation station in Hawaii.

See: Earth Observatory Newsroom, http://earthobservatory. nasa.gov/Newsroom/NewImages/images.php3?img_id=16954

## The twist in the stick

One of the starkest illustrations of the impact of greenhouse gas concentrations on temperature is a graph compiled by the climate scientists Michael Mann, Raymond Bradley and Malcolm Hughes in the late 1990s. The graph shows fluctuations in temperature in the Earth's northern hemisphere from 1000 to 2000. For much of this millennium the graph shows a slight downward trend. This trend changes around 1850 when fossil fuels were being burnt in earnest to power the Industrial Revolution. At this point the graph begins to rise, and continues to rise with ever increasing acceleration. The shape of this graph, with its dramatic upturn in the last section, has led this vivid illustration of temperature change to be called the 'hockey stick' graph.

However, any hockey player tracing the graph's outline may have concerns over the use that something of this shape would be in the game. For in its very last section the 'hockey stick' has an added twist – a brief but even steeper upturn in temperature. Even though the twentieth century warmed quickly in comparison to much of the previous millennium, the last three decades of the century saw an even faster warming – approximately 0.2°C in each decade.[18]

## Goodbye golden toad

So the problem isn't just that the temperature is rising – it is that the rise is taking place at such a phenomenal speed. This makes it difficult for Earth's ecosystems to survive, as there is little time for different species to change their behaviour in response to rapidly changing conditions. Rachel Warren of the UK's Tyndall Centre for Climate Change Research found that ecosystems can really only deal with a temperature increase of 0.05°C per decade.[19] This is much slower than the current global rate of change and is far, far slower than the rate of temperature change in the Arctic. At over 0.4°C per decade this Arctic rate, says Warren, is 'considered sufficient to cause serious ecosystem disruption'[20]. At a 2005 conference on avoiding dangerous climate change, Warren said that:

> effects of [the fairly small climate change that has taken place to date] are already being observed across the world, from the Arctic to the Tropics, from the oceans to the mountains, in the earth system, in ecosystems and in human systems. Across the globe species are changing their phenology and geographical distribution in a direction consistent with their expected response to climate change. Glaciers are melting throughout most of the world, the ocean has already acidified by 0.1 pH units, unprecedented heat waves are causing episodes of mortality in large cities, and drought is intensifying in many regions.[21]

One of the species already pushed off the planet as a result of these changes is the golden toad, which used to live in the cloud forest of Costa Rica but has not been seen since 1989. While it may still be too early to pinpoint the exact cause of the loss of this amphibian, research by Camille Parmesan of the University of Texas claimed that at least 70 species of frogs have become extinct because of climate change.[22] Many of these lived in high mountains so they had nowhere to escape to as temperatures rose. Parmesan found that, around the globe, between 100 and 200 other animal species dependent on cold weather are in trouble.[23] These include polar bears that depend on sea ice as a platform for hunting and a path to coastal areas. 'We are finally seeing species going extinct,' said Parmesan. 'Now we've got the evidence. It's here. It's real. This is not just biologists' intuition. It's what's happening.'[24]

**Table I.1 Possible climate impacts**

| Temp rise | Water | Food | Health | Land | Environment | Abrupt and large-scale impacts |
|---|---|---|---|---|---|---|
| 1°C | Small glaciers in the Andes disappear completely, threatening water supplies for 50 million people. | Modest increases in cereal yields in temperate regions. | At least 300,000 people each year die from climate-related diseases (predominantly diarrhoea, malaria, and malnutrition). Reduction in winter mortality in higher latitudes (Northern Europe, USA). | Permafrost thawing damages buildings and roads in parts of Canada and Russia. | At least 10% of land species possibly facing extinction. Bleaching of 80% of coral reefs, including Great Barrier Reef. | Atlantic thermohaline circulation starts to weaken. |
| 2°C | Potentially 20–30% decrease in water availability in some vulnerable regions, e.g. Southern | Sharp declines in crop yield in tropical regions (5–10% in Africa). | 40–60 million more people exposed to malaria in Africa. | Up to 10 million more people affected by coastal flooding each year. | 15–40% of species possibly facing extinction. High risk of extinction of | Potential for Greenland ice sheet to begin melting irreversibly, accelerating sea level rise and |

| | | | | | |
|---|---|---|---|---|---|
| 2°C | Africa and Mediterranean | | | Arctic species, including polar bear and caribou. | committing world to an eventual 7 m sea level rise. |
| 3°C | In Southern Europe, serious droughts occur once every 10 years. 1–4 billion more people suffer water shortages, while 1–5 billion gain water, which may increase flood risk. | Agricultural yields in higher latitudes likely to peak. 1–3 million more people die from malnutrition (if carbon fertilisation weak). | 1–170 million more people affected by coastal flooding each year. | 20–50% of species possibly facing extinction, including 25–60% mammals, 30–40% birds and 15–70% butterflies in South Africa. Onset of Amazon forest collapse (some models only). | Rising risk of abrupt changes to atmospheric circulations, e.g. the monsoon. Rising risk of collapse of West Antarctic Ice Sheet. Rising risk of collapse of Atlantic thermohaline circulation. |

**Table I.1 continued**

| Temp rise | Water | Food | Health | Land | Environment | Abrupt and large-scale impacts |
|---|---|---|---|---|---|---|
| 4°C | Potentially 30–50% decrease in water availability in Southern Africa and Mediterranean | Agricultural yields decline by 15–35% in Africa, and entire regions out of production (e.g. parts of Australia). | Up to 80 million more people exposed to malaria in Africa. | 7–300 million more people affected by coastal flooding each year. | Loss of around half Arctic tundra. Around half of all the world's nature reserves cannot fulfil objectives. | |
| 5°C | Possible disappearance of large glaciers in Himalayas, affecting one-quarter of China's population and hundreds of millions in India. | Continued increase in ocean acidity, seriously disrupting marine ecosystems and possibly fish stocks. | | Sea level rise threatens small islands, low-lying coastal areas (Florida) and major world cities such as New York, London and Tokyo. | | |

| More than 5°C | The latest science suggests that the Earth's average temperature will rise by even more than 5 or 6°C if emissions continue to grow and positive feedbacks amplify the warming effect of greenhouse gases (e.g. release of carbon dioxide from soils or methane from permafrost). This level of global temperature rise would be equivalent to the amount of warming that occurred between the last ice age and today – and is likely to lead to major disruption and large-scale movement of population. Such 'socially contingent' effects could be catastrophic, but are currently very hard to capture with current models, as temperatures would be so far outside human experience. |

Note: Temperatures represent increases relative to pre-industrial levels. At each temperature, the impacts are expressed for a 1°C band around the central temperature, e.g. 1°C represents the range 0.5–1.5°C. Numbers of people affected at different temperatures assume population and gross domestic product scenarios for the 2080s from the Intergovernmental Panel on Climate Change. Figures generally assume adaptation at the level of an individual or firm, but not economy-wide adaptations due to policy intervention.

Source: Nicholas Stern, *The Economics of Climate Change: The Stern Review*, Cambridge University Press, 2007, p. 57.

## Trouble brewing

While rapid changes in temperature are having a severe impact on habitats, the problem is greater than the fate of each individual ecosystem. For some of the Earth's ecosystems are not simply affected by changes in the temperature – they help adjust the temperature in the first place, through absorbing or reflecting heat. Forests, for example, absorb more heat than open land. They also absorb carbon dioxide, lessening concentrations of this greenhouse gas in the atmosphere. But these functions are now under threat. Mass deforestation has already reduced the amount of land under tree cover. Also, as levels of carbon dioxide in the atmosphere get higher and higher, the ability of tropical forests to absorb the gas gets less and less. High levels of carbon dioxide are expected to lead to high temperatures and the possibility of the rate of respiration (where carbon dioxide is released) overtaking the rate of photosynthesis (where carbon dioxide is absorbed). This would turn forests from carbon sinks, where they absorb carbon dioxide, into carbon sources. A report from the 2005 conference on avoiding dangerous climate change warned that the carbon sink contribution of tropical forests 'appears unlikely to continue for the rest of this century'.[25] It said that: 'plausible mechanisms have been identified which may turn this [ecological community] to a modest or even mega-source of carbon.'

With tropical forests unable to store carbon at rates previously anticipated – and possibly even adding to the amount of carbon dioxide in the atmosphere – climate change may take place at a more rapid rate than formerly predicted. Meanwhile, if forest ecosystems were to collapse and die due to drought or other extreme environmental conditions, their decomposition would lead to a release of the greenhouse gas methane. This would push temperatures up even further:

although methane stays in the atmosphere for a shorter time than carbon dioxide, it is 24 times as potent a greenhouse gas.

Like forests, the Earth's vast oceans have acted as a natural reservoir of carbon dioxide. They have absorbed up to half of the carbon dioxide from fossil fuel emissions and other human activities in the last 200 years.[26] While this may have slowed down changes in the climate, the condition of the oceans themselves has changed during this time, impacting on their capacity to continue to absorb more of the gas. As carbon dioxide reacts with seawater the ocean has become more acidic. Researchers at the Marine Laboratory in Plymouth, UK, found that while the average acidity of seawater changed by less than 0.1 unit over the several million years before the start of the Industrial Revolution, since then the acidity of the ocean's surface waters has already changed by 0.1 unit.[27] They concluded that: 'the marine system is accelerating its entry into uncharted territory.'[28]

The increasing acidity of the ocean limits the development of marine organisms' shells and skeletons – structures made of calcium carbonate which, until now, have been one of the main mechanisms by which carbon was locked away and eventually buried on the ocean floor. Some of these tiny marine organisms are major producers of dimethyl sulphide (DMS) which may help regulate climate through the production of cloud condensation nuclei.[29] A reduced flow of DMS from the oceans to the atmosphere could mean further increases in global temperatures due to cloud changes.

Another great climate regulator to date has been the gleaming ice surrounding the Earth's poles. This too is under threat. In August 2006 the European Space Agency noted the appearance of 'dramatic openings' in the Arctic sea ice, which is usually frozen all year.[30] Ice sheets have played an essential role in cooling the planet, reflecting light and heat away from the Earth. As ice sheets melt, there is less shiny white surface area to deflect this light and heat, which are absorbed instead by the darker seas. Less ice in one year makes it more difficult for ice to re-form the following year. The decline of the ice sheets – and the demise of this temperature regulator – thus accelerate. The faster the sheets melt, the faster they will continue to melt. Already, newly formed lakes and cracks in ice are making the Arctic disintegrate faster than expected. Until fairly recently it was thought that it would take up to 1000 years for heat to penetrate and melt the Greenland ice sheet. But research in February 2006 in the journal *Science* showed that the amount of ice dumped by glaciers into the sea had almost doubled in the previous five years.[31] The data suggested that Greenland was losing at least 200 cubic kilometres of ice a year, whereas only two years previously the ice sheet was still thought to be in balance.[32]

This melting ice contributes to sea level rise, which endangers lives and will displace millions of people. John Young, former senior researcher at the Worldwatch Institute, has said that one of the greatest costs of rising sea levels will not be the loss of land but the inevitable disruption of communities and cultures that cannot be replicated elsewhere.[33] For Africa alone, a 2006 UN report found that the number of people at risk from coastal flooding due to sea level rise will rise from one million in 1990 to 70 million by 2080.[34] As much as 30 per cent of Africa's coastal infrastructure – including the cities of Alexandria, Lagos, Cape Town, Maputo and Dar Es-Salaam – could be flooded.[35]

A melting Arctic will release the greenhouse gas methane that lies in the permanently frozen subsoil – known as permafrost – beneath the Arctic's shallow seabed and in the floodplains of Arctic lowland rivers. This permafrost is now releasing methane five times faster than expected.[36] Lakes are appearing on the surface of previously frozen peat bogs in Siberia, where the permafrost can run up to a mile deep.

Another study published in *Science* found that, alongside the trapped methane, the amount of carbon dioxide trapped in this type of permafrost may be 100 times the amount of carbon released into the air each year by the burning of fossil fuels.[37] 'It's kind of like a slow-motion time bomb,' said the study's co-author Ted Schuur.[38]

## Over we go

Like changes in the forests and oceans, changes in ice sheets not only undermine the role of ecosystems in regulating climate but may also trigger a release of greenhouse gases that would amplify changes in the climate and send them accelerating beyond control. Writer Fred Pearce describes such a scenario as 'an atmospheric tsunami, swamping the planet in warmth'.[39]

In his book *The Last Generation*, Pearce describes how ice cores extracted in Greenland and Antarctica reveal that such an abrupt change to the climate system is not unprecedented. An ice core is a sample drilled out from layers of snow and ice that have built up over many years, used in studies by organisations such as the European Project for Ice Coring in Antarctica (EPICA). The cores contain tiny air bubbles, as well as wind-blown dust and ash that were trapped as the snow and ice formed. All these can add up to form a picture of the climate – the chemistry and gas composition of the atmosphere, the history of volcanic eruptions and solar activity – over hundreds of thousands of years. One EPICA core drilled in Antarctica went back 720,000 years and revealed eight previous glacial cycles. The findings from some polar ice cores suggest that around 12,000 years ago the warming that was lifting the Earth out of an ice age went into an abrupt reversal.[40] The ice age returned for another thousand years before ending with such speed that the world warmed by at least five degrees within ten years. The ice cores demonstrate a close link between greenhouse gas concentrations and such temperature changes. They raise the question: given that today's greenhouse gas concentrations are at an unprecedented high, are we pushing ourselves to a point where we tip over into a world of sudden and unpredictable climate events?

Looking at the current behaviour of the world's ice sheets, James Hansen, director of NASA's Goddard Institute for Space Studies, noted that once a sheet starts to disintegrate it can reach a point beyond which break-up is explosively rapid and the ecosystem of which it is part flips into a different state. 'How far can it go?' asks Hansen.[41] Other individual parts of the climate system raise the same question. How close are we to massive releases of greenhouse gases as tropical rainforests shift from carbon sink to source, or the Siberian peat bogs turn to slush? How much longer until the glaciers that feed water supplies for China and India hit irreversible meltdown? While we don't know the precise answers to these questions, or the effects of passing each of these individual 'tipping points', we must accept that the direction in which we are currently going – towards them – has to change.

## Brave new world?

Joanna, whose story we heard at the beginning of this book, may not yet have reached a tipping point into impossible living conditions but she is certainly struggling to cope. Moving from the scale of an individual to an entire nation, the low-lying island state of Kiribati has already been tipped into almost impossible living conditions as it is highly vulnerable to sea level rise. The island's residents are beginning to leave: about 17,000 islanders applied for residence in New Zealand in 2004 and 2005, compared with 4000 in 2003.[42] 'For Kiribati the tipping point has already occurred,' says Stephen

H. Schneider of Stanford University. 'As far as they are concerned, it's tipped, but they have no economic clout in the world.'[43]

The purpose of this book is to show what those with little economic clout have been saying for some time: we live in a world where rising temperatures are affecting the path of – and potential for – meaningful human development, so we have to do something to stabilise the climate. Obviously 'doing something' must include reducing global greenhouse gas emissions and developing less polluting energy systems. However, it also means adapting to cope with the changes already locked into the climate system, and ensuring that adaptation in poorer countries is beneficial and is fully supported by richer ones. As such, this book is a call for climate justice: the world's rich are responsible for the bulk of past emissions, yet the world's poor, who have least resources to adapt, are most vulnerable to their effects. Whether rich or poor, we all have the right to long and fulfilling lives within a stable climate.

## Box I.2: Choices

Choices made by individuals, institutions and governments over the next few decades will determine the degree and nature of future climatic and environmental change for millennia to come. Twenty-first century choices regarding energy technologies, economic models, intellectual property and technology transfer, equity and sustainable development, and patterns of consumption will determine the exposure and vulnerability of future generations. ... The future nature of our world, globalised or otherwise, will depend on how these processes are managed. Global environmental change will be a fact of life for future generations, and it will alter their exposure to environmental risk and hence partially determine their vulnerability to natural disasters. The extent of the change with which they have to cope, and the options available for minimising their vulnerability, will be decided by the politics of today.

W. Neil Adger and Nick Brooks, 'Does global environmental change cause vulnerability to disaster?', in M. Pelling (ed.) *Natural Disaster and Development in a Globalising World*, Routledge, 2003.

# 1
# Preparing to be prepared

> The political challenge of climate change is one of empowerment, to ensure that the great majority of the world's population are able to evolve as climate continually alters, rather than cope with ever-changing stress – a world of winners rather than of victims and losers.
>
> Adger, Brooks, Bentham, Agnew and Eriksen, 'New indicators of vulnerability and adaptive capacity'[1]

## Separated at birth: mitigation and adaptation

So far, much of the international discussion on tackling climate change has focused on mitigation – limiting emissions in order to lessen the severity of the climate crisis. This makes a lot of sense, as the root of a problem needs to be addressed if it is to be tackled effectively. But even if we were to reduce greenhouse gas emissions now, some degree of climate change is inevitable and unstoppable. Sea levels are rising and will continue to do so for centuries due to the surface warming of the oceans that has already occurred.[2] Meanwhile, carbon dioxide stays in the atmosphere for around a century so greenhouse gas concentrations will also continue to rise. The climate we live with today is influenced by the emissions of a previous generation. The emissions we produce today are defining the climate for at least the next generation.

The science academies' statement that was released to coincide with the 2005 G8 meeting said the following:

> Even if greenhouse gas emissions were stabilised
> instantly at today's levels, the climate would still
> continue to change as it adapts to the increased emis-
> sions of recent decades. Further changes in climate are
> therefore unavoidable. Nations must prepare for them.[3]

This preparation must include something that the poorer
countries of the world have been calling for for some time –
adaptation. Adaptation means reducing vulnerability to
changes in the climate. The UK's Tyndall Centre for Climate
Change Research defines it as the adjustment of a system to
moderate the impacts of climate change, to take advantage
of new opportunities or to cope with the consequences.[4]
Adaptation involves looking further and deeper than the
predicted impacts of climate change. For example, the
researchers Lisa Schipper and Mark Pelling looked at the
situation in El Salvador and found that:

> In the case of agriculture, a focus on impacts would
> examine changes in crop yields resulting from more or
> less rainfall and the ways in which societies can adapt
> to these varying yields. It would not address related and
> dependent issues of unemployment, poverty, lack of
> technical support and food security.[5]

In aiming to address wider issues, such as the impact of
climate change on poverty and food security, adaptation
shares many aims with existing development programmes.
Taking the example of Bangladesh, Dr Saleem Huq of the
International Institute for Environment and Development and
Mizan Khan of the North South University in Bangladesh
wrote: 'adaptation is an option not by choice, but by compul-
sion, as insurance to [Bangladesh's] efforts in achieving
sustainable development'.[6]

## Adaptive capacity

The precise nature of the changes that will take place in our warming world is difficult to pin down, particularly at a very local level. This does not mean doing nothing until changes take place and their nature becomes clear. Far from it. The unpredictability of the whole situation means that strengthening a general capacity to adapt – preparing, as it were, to be prepared – is essential.

This capacity to adapt will depend on several factors. The ADAPTIVE project, based at the Universities of Oxford and Sheffield, UK, looked at how adaptations to climate change could work for societies in poorer countries that were dependent on natural resources. The project found that the communities most able to cope 'are those which are most cooperative and with the strongest social institutions'.[7] Such communities 'are able to innovate and experiment in the face of change, as well as drawing on traditional knowledge and networks'.[8] In Nwadja-hane in Mozambique (the area where Joanna, who featured in the Introduction, lives), people reduced their vulnerability to climate change by taking collective action. For example, one way in which farmers tried to adapt to climate change was to set up new farming associations that could spread the risk of new practices and technologies.[9] Their experience of having worked in partnership before formed a useful background for this. '[People] are not helpless in the face of these major changes,' said the ADAPTIVE project's Professor David Thomas.[10]

The importance of social institutions in strengthening adaptive capacity goes beyond the local level. Governing institutions that manage resources on a wider level also have a role to play. For example, a person with good healthcare and a good education as a result of government investment in these areas has a better chance of adapting to cope than someone who is struggling simply to survive.

Poverty clearly has an impact on adaptive capacity as it means less access to resources. Poorer people may be more vulnerable to climate change simply because of where they live. Poorer groups are more likely to live in exposed areas such as flood plains or unstable hill slopes – the places that anyone with resources can choose to avoid.[11] Poor and low-quality housing is more vulnerable to floods or storms. Poorer communities are less likely to afford insurance to deal with any climate change problems that do occur, and have limited access to alternative sources of food and income when the main source fails.

Poverty means fewer resources to make adaptations, even when there is awareness that climate change is going to have an impact on homes and livelihoods. One project in Zimbabwe explored whether an awareness of seasonal climate forecasts would help subsistence farmers. Researchers found that nearly half the participants said that even if they wanted to make adjustments based on the forecasts, 'shortage of money and credit' would prevent them from doing so.[12] Similarly, Inuit interviewed for the case study in Box 1.1 (see below) were aware that different equipment would help them cope with the changing conditions. Unfortunately, they did not always have the money to buy it.

## Box 1.1: Community resilience in the Arctic

If [Inuit] weren't adaptable they wouldn't be around.
John MacDonald, Igloolik*

A 2005 report by the Global Environmental Change Group at the University of Guelph looked at how two Inuit communities in Nunavut in Canada are vulnerable to climate change. The report found recent climatic changes that are unusual, indeed in many cases unprece-dented. Increased weather and ice unpredictability, and stronger winds during the break-up of sea ice have caused problems, particularly for harvesting activities, which have become more dangerous. Sudden shifts in the wind have stranded hunters on drifting ice; in 2000, 52 hunters had to be rescued by helicopter, and many lost their equipment.

In response, people are becoming more risk averse, avoiding travelling if they believe the weather is going to be bad, returning quickly if they are out on the land when weather conditions change, and generally being more vigilant. Many hunters take extra supplies or take rowing boats on their sleds in case the ice breaks up. Their equipment has also been modified. For example, boats with more powerful outboard engines allow hunters to spend less time on exposed water.

The resilience demonstrated by the two communities in the study is facilitated by traditional knowledge and behaviour codes known as *Inuit Qaujimajatuqangit* (IQ). IQ involves strong links among family and friends. Safety equipment is shared between friends and family, as hunt-ing equipment has been shared for generations. Information on changing conditions and risks is also freely passed on, with elders going on the radio to tell people what to do if they get into danger.

Learning this IQ requires being on the land regularly, and observing others. Yet younger people are less keen to do this and the IQ that has enabled Inuit to live and thrive in the unpredictable climate of the Arctic is being eroded, reducing the communities' ability to adapt.

Source: 'Living with change in Nunavut: vulnerability of two Inuit communities to risks associated with climate change',

James Ford, Barry Smit, Johanna Wandel, University of Guelph, presentation to the conference on *Adapting to Climate Change in Canada 2005: Understanding Risks and Building Capacity*, Montreal, 2005.
\* Ibid, p. 10.

The low-lying countries of the Netherlands and Bangladesh are often used to illustrate how poverty can determine adaptive capacity at a national level. Both countries are vulnerable to sea level rise but one is rich, the other poor. Their adaptive

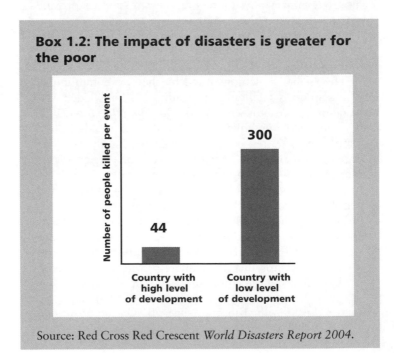

**Box 1.2: The impact of disasters is greater for the poor**

Source: Red Cross Red Crescent *World Disasters Report 2004*.

A 2006 World Bank report found that the incidence of disasters increased from fewer than 100 in 1975 to more than 400 in 2005. The report's author Ronald Parker said he believed this was 'clearly tied to environmental degradation around the world' (*Peopleandplanet.net*, 24 April 2006).

capacity varies accordingly. If the best solution for both countries was to build sea walls, then the Netherlands has the technological, social and financial capital to do this while Bangladesh does not.[13]

## Box 1.3: Adaptations in Pakistan

Interview by the author with Hafeez Buzdar, Programme Manager for Sustainable Livelihood & Disaster Preparedness, Oxfam's Pakistan Programme, 17 April 2006.

MJ: What changes in the climate and environment are the small-scale producers that you work with experiencing?

HB: They are experiencing increased poverty and vulnerability due to drought in Balochistan, floods, sea intrusion and reduction in mangrove forests in coastal Sindh, and floods in southern Punjab. Floods and cyclones also adversely affect the fishing communities of coastal Sindh and they lose their lives, boats, fishing nets, huts and livestock. Meanwhile, cultivated land is becoming increasingly saline and degraded due to successive floods and sea intrusion, and there is a shortage of drinking water in the coastal fishing communities as well as in desert areas of Sindh and Balochistan Province.

MJ:   How are people responding to these changes?

HB:   They are raising 'people bargaining power' through local level institution building and capacity building; they are developing early warning systems and a local disaster plan, and developing the infrastructure for water-harvesting. Other actions include implementation of a Farmers' Information, Services and Demonstration Centre for capacity-building and awareness; rehabilitating degraded lands in Lower Sindh through biological drainage (wide-scale tree plantation); promoting sustainable agricultural practices and other low-cost appropriate technologies, including the use of indigenous environmentally-friendly pesticides and water-saving irrigation methods. We are planning to conduct a study on water policy and management in Balochistan in order to have a more informed position. Finally, we will carry out policy advocacy with government agencies on sustainable agronomic practices.

MJ:   What factors influence the ability to manage climate change?

HB:   Awareness and understanding of the issues; community participation; effective public institutions and strong governance; sustainable water management policy initiatives, for example withdrawing subsidies for the electricity use in mining, overuse and exploitation of groundwater in Balochistan; increase in forest and vegetation cover; and policy advocacy for climate change adaptation efforts.

MJ: What factors limit the ability to adapt?

HB: Lack of understanding of the consequences; lack of resources; population growth; short-term faulty incentives; lack of enforcement of policy, conventions, rules and regulations; absolute poverty.

## Making matters worse

Tackling poverty, then, is essential if people are to have the capacity to adapt in a changing climate. Yet climate change itself may actually undermine, and even reverse, international efforts to do something about poverty. These efforts come in the form of the United Nations' Millennium Development Goals (MDGs), launched in September 2000 to be, in the words of former UN Secretary-General Kofi Annan, 'a blueprint for building a better world in the twenty-first century'.[14]

Climate change will impact on these Goals in the ways shown below.

### *Millennium Development Goal 1: eradicate extreme poverty and hunger*

Brian Halweil, a senior researcher at the Worldwatch Institute described farming (the basis of food production) as 'the human endeavour most dependent on a stable climate – and the industry that will struggle most to cope with more erratic weather, severe storms, and shifts in growing season lengths. While some optimists are predicting longer growing seasons and more abundant harvests as the climate warms, farmers are mostly reaping surprises.'[15] It is true that small increases in temperature can make some crops more productive and can mean that crops grow at higher altitudes than before. However, the overall impact of climate change on agriculture will be negative. If the average global surface temperature rises by 2°C above pre-industrial levels, predictions for the loss in global cereal production range from 30 to 180 million tonnes.[16] Between 3.3 and 5.5 billion people are living in countries or regions expected to experience large losses in crop production potential if global temperatures rise by 3°C[17] – an increase likely this century without urgent action to limit emissions. This spells disaster for poor communities in particular, who tend to be more dependent on agricultural practices. About 70 per cent of Africa's total population and nearly 90 per cent of its poor primarily work in agriculture.[18] Rainfall, which is increasingly unpredictable as a result of climate change, is the only source of water for much of Africa's subsistence farming.

The number of extra people at risk from hunger by the 2080s is predicted to be 45 to 55 million for a temperature increase of 2.5°C.[19] For an increase of 3°C, 65 to 75 million extra people may be at risk, and up to 80 to 125 million may be at risk for an increase of 3 to 4°C.[20]

### Millennium Development Goal 2: achieve universal primary education

In an article for *Tiempo* magazine, Hannah Reid and Mozaharul Alam described how climate change would affect this second Millennium Development Goal. They said that natural disasters and drought may require children to help more with household tasks, leaving less time for schooling.[21] School buildings themselves are under threat from weather-related disasters – a quarter of Honduras's schools were destroyed during Hurricane Mitch in 1998.[22] Increased malnutrition (see Goal 1) or disease (see Goals 4, 5 and 6) may make it less likely that a child attends school. School attendance also becomes a low priority when weather-related disasters lead to the loss of a home, or a need to migrate.[23]

### Millennium Development Goal 3: promote gender equality and empower women

About 70 per cent of people in the developing world who live below the poverty line are women.[24] While this suggests that women are already an extremely vulnerable group, the nature of their livelihoods makes them even more so. Women are often responsible for water collection – a task that will become harder as climate change affects availability (see Goal 7). Women are also often responsible for the cultivation and production of farm crops, with women making up almost 80 per cent of the agricultural workforce in Africa.[25] Agriculture is highly vulnerable to climate change (see Goal 1). Women employed in specific agricultural work, such as paddy cultivation in Asia, may find that their jobs, like the industries, are under threat as a result of climate change.[26]

A 2004 report on the threats from, and responses to, the impact of global warming on human development warned that: 'climate change policies will be unsuccessful if women have no opportunity to influence decision-making, build their capacity, lower their vulnerability, and diversify their income sources.'[27]

### Millennium Development Goals 4, 5 and 6: health related issues (reduce child mortality; improve maternal mortality; combat HIV/AIDS, malaria and other diseases)

According to the World Health Organisation, 'the long-term good health of populations depends on the continued stability and functioning of the biosphere's ecological, physical, and socio-economic systems.'[28] Such stability is threatened by climate change. Likely health problems range from direct effects such as heat stress and weather disasters to increased occurrence and distribution of vector-borne diseases such as dengue fever.

Health problems from climate change include the impact of hunger and malnutrition (see Goal 1). They include the effects of both too little water, as people are pushed to drink from ever more unsafe water sources, and too much water, as floods and storm surges contaminate drinking water supplies (see Goal 7). The impact of climate change on biodiversity (see Goal 7) will also impact on people's ability to cope with illness, as approximately 80 per cent of the population of poorer countries rely on plants as traditional medicines for primary health care.[29]

## Box 1.4: Vulnerability to dengue fever in Jamaica

Researchers in Jamaica found that the country's Ministry of Health was aware that climate change in the Caribbean could produce increasing temperatures and precipitation and that these changes could increase health problems in the form of dengue transmission. Yet in spite of this, no long-term strategies to cushion possible negative impacts had been put in place or were even under consideration. Dengue fever was given significantly less priority than some other diseases, especially HIV/AIDS. This may have been because of inadequate funding and a need to establish priorities.

Researchers found a view in the Ministry of Health that communities must take responsibility for control of disease carriers. The project researchers agreed with this but emphasised that this has to be a policy position rather than a defensive posture, and must be supported by initiatives aimed at empowering communities to assume control. Public education, for example, was seen as necessary to address knowledge gaps over what causes the disease, the nature of its symptoms, and the risks associated with improper water storage.

The project researchers also found that public sector organisations with the mandate to mitigate hazards and promote sustainable development seemed obsessed with the threats posed by sea level rise. 'No one can deny the threat posed by this phenomenon,' they said, concluding:

> But it is difficult to see why all of the public sector agencies should become so absorbed with the gradual encroaching of the sea to the exclusion of more imminent threats. They should be persuaded to broaden their concept of a hazard; to realise that the threat of an increase in the occurrence of a debilitating and possibly deadly disease is not incompatible with their mandate; to see threats to health as threats to sustainable development and include these issues in their public education programs. There is need for a concerted effort of collaboration with various public and private sector environmental organisations. These are elements in the country's generic capacity which constitute the foundation for adaptation.

From: C. Heslop-Thomas, W. Bailey, D. Amarakoon, A. Chen, S. Rawlins, D. Chadee, R. Crosbourne, A. Owino, K. Polsom, *Vulnerability to Dengue Fever in Jamaica*, Assessments of Impacts and Adaptations to Climate Change (AIACC) Working Paper No. 27, May 2006.

## Box 1.5: Disease and climate change

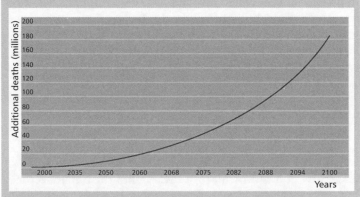

Predicted extra deaths from disease attributable to climate change in sub-Saharan Africa based on the scenario of a 6°C temperature rise by the end of the century.

Christian Aid produced this graph and pointed out that these figures are a projection, so cannot be absolutely precise, but they do indicate the vast scale of the problem.

Source: Christian Aid, *The Climate of Poverty: Facts, Fears and Hope*, 2006, p. 9 (http://www.christian-aid.org.uk/indepth/605 caweek/caw06final.pdf).

### *Millennium Development Goal 7: ensure environmental sustainability*

One of the greatest challenges posed by climate change is the threat to biodiversity. A 2004 report in the journal *Nature* found that 15 to 37 per cent of species in sample regions covering about 20 per cent of the world's land area will be 'committed to extinction' as a result of mid-range climate warming scenarios for 2050.[30] Already, coral reefs, a building block for marine life, have suffered devastating losses as a result of increased water temperatures. For those species that survive any rise in temperature, climate change will mean a change in the boundaries of their ecosystems. Species that live in mountainous, coastal or other geographically restricted areas will have few alternative habitats to which they can migrate. Unpredictable weather will also have impacts on reproductive cycles, growing seasons and food webs, including relationships between predators and their prey.

The UN-supported Millennium Ecosystem Assessment found that those who lack minimum standards of human well-being are also those 'most vulnerable to the deterioration of natural systems'.[31] In Namibia, where over 30 per cent of gross domestic product relies on natural resource-based production like agriculture, fisheries and mining,[32] the Ministry of Environment and Tourism said:

> Climate change is one of the most serious threats to Namibia's environment, human health and well-being as well as its economic development. The arid environment, recurrent drought and desertification have contributed to make Namibia one of the most vulnerable countries to the effects of climate change. Considering the natural resource based economy and limited technical and financial resources, climate change could potentially become one of the most significant and costly issues that affect the national development process in Namibia.[33]

The seventh Millennium Development Goal includes a key component for development – the target to halve the number of people without access to safe drinking water. Water supplies are expected to decrease as weather patterns become more extreme. In some parts of Africa access to water has become highly unreliable, with floods and droughts occurring within months of each other. In India, studies on rainfall data from central India from 1951 to 2000 found worse rainstorms and a greater risk of floods.[34] 'Extreme monsoon rainfall events' have become more common but the 'moderate rainfall events' decreased.[35] This is a crucial change because moderate rainfall provides a steady supply of water seeping into the ground. Without sufficient storage capacity, flooding or heavier rainfall does not necessarily mean an increased water supply. Heavier rainfall in short bursts may mean that crops receive the same amount overall but they can suffer shortages between storms and, when too much falls, much is wasted.[36]

In Asia, the retreat of glaciers as ice melts under increasing temperatures means reduced water supplies to rivers, which means reduced supplies for agriculture and for drinking. Oxfam's John Magrath wrote that: 'A decrease in water supply due to glacier melt is likely to exacerbate droughts, increase competition and possible conflict.'[37]

The rise of sea levels as a result of climate change may also mean that saline water enters groundwater aquifers near coasts, decreasing water resources.

If the average global surface temperature rises by 2°C above pre-industrial levels, predictions for the number of people experiencing an increase in water stress range from 1.0 to 2.8 billion.[38]

## Millennium Development Goal 8: develop a global partnership for development

According to the new economics foundation:

> Aid to ensure that poor countries are better prepared to deal with weather-related disasters is important, but, in

the long run, will not be enough. More importantly there needs to be frameworks to combat the causes of climate change, with particular sensitivity to the world's poor.[39]

### *The challenge presented by the Goals*

A 2004 report on threats from, and responses to, the impact of global warming on human development said:

> Today, humanity faces the intertwined challenges of obscene levels of poverty and a rapidly warming global climate. There is no either/or approach possible; the world must meet both its commitments to achieve the Millennium Development Goals and tackle climate change.[40]

To emphasise the impact that climate change is having on poverty, Christian Aid suggests that a ninth Millennium Development Goal – a call for governments to reduce emissions – should be added to the existing eight.

## Communicating on climate change

The two issues of poverty and climate change are not always seen as inextricably linked. Climate risks may be seen as far off or uncertain, and therefore less important than the need to use already limited resources to meet basic day to day needs, or to deal with immediate crises such as HIV/AIDS. Meanwhile, some impacts of climate change – droughts and storms for example – may just seem like more of a problem that already exists.

Some of the lack of concern over climate change may come from limited information on the subject. In the case study in

Box 1.4 (Vulnerability to dengue fever in Jamaica), researchers found a lack of awareness of climate change, although this failure had not happened by default. A spokesperson at Jamaica's Office of Disaster Preparedness and Emergency Management told researchers that although the agency felt it ought to be communicating about climate change to the public, it had not included the subject in its public education campaign because:

> climate change is not a simple topic to grasp. The average man cannot relate to it as it seems very far-fetched to him. ... Its effects are long term and ... individuals don't care about things that do not affect them immediately.[41]

The Office might have looked for ways to communicate information about climate change, but instead the issue had been avoided altogether!

In contrast, Pablo Suarez of Boston University has researched how best to convince communities at risk about the need to invest in long-term climate change planning, such as adaptation. Suarez found that the seed of addressing the threat of increased climate hazards may lie inside, rather than counter to, the experience of facing urgent and tangible problems such as lack of drinking water, sanitation, health services or food security. 'Successful disaster risk reduction initiatives,' says Suarez, 'usually require a level of community involvement that can only emerge from shared understanding of past and future threats.'[42] He believes that in addition to mainstreaming long-term adaptation into existing, more urgent activities, the reverse process can be explored. Instead of focusing solely on the future, climate information can be used to strengthen the case for present action. Suarez worked with subsistence farmers in Mozambique, who tend to explain floods or droughts through supernatural intervention – as punishment from God, an expression of their ancestors' anger, or bad luck. As a consequence, the farmers think that these events, which have already taken place, are unlikely to happen again. This reduces their willingness to take adaptive measures such as setting up early warning systems or planting drought-resistant crops. Suarez suggests that climate change information can help this group of farmers see that those extreme events may be related to a different, global, causal process, and are therefore more likely to occur again. 'Climate change information and resources', says Suarez, 'can strengthen existing disaster management projects, empowering communities to improve their short-term horizon and consequently be more capable to adapt to a different climate.'[43]

## Box 1.6: Communicating climate information at a community level

Helping people to understand, interpret and use complicated scientific and technology-based knowledge is at the very core of economic development in the twenty-first century. Successful methods for communicating information include the principles outlined below.

### *Dialogue, not monologue*

Climate information tends to be communicated unidirectionally. When forecasts are 'dumped' on communities, the information is unlikely to lead people to more informed decisions because it is either not understood or not believed, or because the listeners do not know what to do about it. The process of communication needs to allow users to engage in a dialogue, both with the source of information (for example, people who understand the forecast well and can clarify any doubts from the potential users) and within the community (for example, subsistence farmers discussing alternative ways to respond to the forecast individually, as a community, or with outside help).

### Keep it simple but honest

Climate information tends to be communicated in an excessively complex manner, and with insufficient acknowledgement of uncertainties. Community-level users should not have to learn too many new words, or receive a forecast with common words that in meteorology have different meanings from the common language. The use of analogies and metaphors can greatly enhance understanding of climate processes. Whenever possible, have a local person repackage the information in order to tailor it to the community's communication culture.

It is indispensable to highlight the probabilistic nature of climate forecasts; community users understand that all knowledge may be limited, and do appreciate honesty in the alleged reliability of predictions. Explaining in simple terms the underlying causes of uncertainty can greatly enhance trust in the forecast.

### Address diversity

Climate information tends to be packaged in ways that assume a rather homogeneous audience. Differences across and within communities need to be fully acknowledged and addressed while designing the contents and the communication process, particularly with regards to response strategies. Relevant aspects of diversity include relative wealth, gender, and cultural willingness to change (often correlated with age).

Source: email from Pablo Suarez, 14 April 2006.

## Influencing adaptation

The November 2001 meeting of the parties to the United Nations Framework Convention on Climate Change recognised that the world's poorest countries are the most vulnerable to climate change. This international forum called on the world's 49 Least Developed Countries to develop national adaptation programmes of action, known as NAPAs. NAPAs are action-oriented programmes geared towards urgent, immediate needs to strengthen a country's capacity to adapt. Samoa's NAPA, for example, includes projects to tackle community water resources and prevent forest fires. Given that much of Samoa's population lives within a kilometre of the coastline, the NAPA also includes a project to manage the coastal infrastructure of highly vulnerable districts.

The preparation of NAPAs is meant to be guided by 'a participatory process, involving stakeholders, particularly local communities'.[44] The extent to which this happens depends on the nature of the implementing agencies within each country. This varies. Bhutan, for example, found an initial hurdle in the fact that most experts on its NAPA team were dealing with climate change issues for the first time, and lacked adequate awareness of climate change and its impacts.[45] Those responsible for guiding the process had their own capacity issues to deal with, before even getting to issues around climate information at a more local level, where awareness was likely to be even less.

Another problem with NAPAs is that, although they are part of the international climate change process, funding to actually implement them has not been forthcoming. This funding was supposed to come from a Least Developed Countries Fund, set up in 2001 to take voluntary contributions from the world's richer countries. By April 2006 there was $89 million in the pot, of which $12 million was to be used to develop the NAPAs, leaving only $77 million for implementing adaptation measures suggested in them.[46] To put this figure in perspective, a report commissioned by the UK government on the economics of climate change found that the cost of building 8000km of river dykes in Bangladesh is $10billion.[47]

Mohammad Reazzuddin, director of the department of the environment in Bangladesh, said at a climate meeting in Nairobi that, on its own, his country could probably identify $100 million worth of urgent and immediate projects.[48] He said that every other country could probably do the same. Yet in the first round of NAPA funding, each country will get no more than one project funded. Even if NAPA funds rise to about $150 million this means none of the Least Developed Countries will get more than $3 million. If Bangladesh wants more – for example, $6 million for urgent immediate needs – it has to raise half the money from the banks. Reazzuddin told Oxfam: 'It's peanuts. It will not deliver the real needs.'[49]

Not only do those countries and communities that are most vulnerable to climate change have limited resources to implement adaptation measures, they also have limited political power to press for greater funding to be made available. And even though it is unlikely that enough funds will ever be available to fully implement NAPAs, there is currently no clear process for prioritising spending – funding food production, for example, over flood protection, or finding a balance between proactive and reactive responses to climate change. While NAPAs may be an important start they are, as yet, not enough to address adaptation needs.

## Keeping up the pressure

Even if adaptation does go some way towards protecting communities from changes in the climate, there is still a great need to reduce emissions. There are, after all, limits to what adaptation can achieve. Adaptation becomes less effective as greenhouse gas concentrations in the atmosphere – and so climate instability – increase. Adaptation needs support, but as a complement to, not a replacement for, emissions reductions.

### Box 1.7: Dear Sir/Madam ...

... hundreds of millions of the world's poorest people [live] in developing countries where adaptation to climate change is ... a life and death issue. ... The severe drought that swept Kenya and much of Ethiopia in 2006 not only exposed some 11 million people to the threat of starvation, it also wiped out assets, increased malnutrition and forced parents to withdraw their children from schools. These are outcomes that have lifelong consequences, creating cycles of vulnerability [and] disadvantage that keep people in poverty across generations. In Malawi, climate change could lead to productivity losses for maize, the national food staple, of more than 20 per cent, according to climate change models. This is in a country where almost two-thirds of the rural population lives below the poverty line, and where half of all children are stunted. ... Failure to put climate change adaptation in poor countries at the heart of the multilateral agenda on climate change will have grave consequences. In many countries it will slow, or even reverse, human development, exacerbating deep national and global inequalities in the process. We know the scale of the threat. It is time the rich world faced up to its responsibility to act.

Source: Letter from Kevin Watkins, Director, Human Development Report Office, United Nations Development Programme to the *Financial Times*, 7 February 2007.

# 2
# False solutions

We are not only victims of climate change, we are now victims of the carbon market.

Jocelyn Therese, Coica (the co-ordinating body of indigenous organisations of the Amazon basin)[1]

## Introducing the Clean Development Mechanism

Chapter 1 showed how adaptations in people's lives are taking place because the world is warming. This chapter will show how changes are taking place because of attempts to tackle the warming. These changes are different in form and motivation to the adaptations outlined previously, but they still take place in response to climate change. They illustrate yet another way in which climate change is having a profound influence over the way that people live, and a profound influence over the direction of development.

Muzelli, whose story featured in the Introduction, has undergone changes – adjustments made to his house to offset emissions from the G8 politicians. The idea behind Muzelli's home improvements underpins a central tool in the Kyoto Protocol (the international agreement on climate change). This tool is the Clean Development Mechanism, known as the CDM. The CDM allows countries that are struggling to meet their Kyoto targets through reductions at home to take a different approach and buy emissions credits from poorer countries. The credits come from investments in low-emission or energy demand-reducing projects. According to the World Bank, the logic of the CDM is that: 'greenhouse gases mix uniformly in the atmosphere, which makes it possible to reduce carbon emissions at any point on the Earth and have the same effect.'[2] The amount of credits earned in any CDM deal is based on the difference between (1) the amount of greenhouse gases that would have been emitted if the CDM project host had followed a more traditional development path (the 'baseline'), and (2) the amount of greenhouse gases generated when a country follows a low-emission path.

For example, Italy and Finland edged closer to their Kyoto targets by buying credits from hydropower dam projects in Honduras. The dams are supposed to emit less greenhouse gases than fossil fuel-based stations that produce equivalent electrical power.

Using credits from elsewhere to meet Kyoto emissions targets is attractive to richer countries because it allows politicians to avoid more costly changes to domestic behaviour patterns. However, while CDM-style emissions reductions avoid big behaviour changes by richer countries, they do influence the behaviour of those who host CDM projects. In Muzelli's case, should he get the urge to install a glass ceiling, tarpaulin or flimsy skylight atop the four walls of his home, he may not be allowed to do so without consulting project managers, as his roof is vital to the emissions accounts of world leaders.

This defining of behaviour is an intentional part of the CDM, which is meant to be more than an accounting tool. While its first aim is to support richer countries in reaching their greenhouse gas reduction targets, its second aim is to support sustainable development in poorer countries. Richard Kinley, former acting head of the United Nations Climate

Change Secretariat, described this as 'helping the developing nations to improve the quality of life for their citizens while also allowing developed nations to earn emission allowances'.[3] While this win–win situation – reducing emissions while supporting a more sustainable development path – sounds like it has potential, the CDM is not working quite so well in practice. The mechanism's two aims appear to fundamentally conflict. This raises a question: are these behaviour changes made under the CDM to accommodate the emissions of richer countries really worthwhile?

## Having it both ways?

Two case studies in the Philippines attempted to meet both of the CDM's objectives.[4] In the first study, small hydropower systems were installed in remote villages that were not connected to the electricity grid. These systems were intended to provide lighting for homes and village centres, and to power simple tools for livelihood activities. The second study involved the hypothetical installation of solar-powered systems in a remote rural village not reached by the electricity grid in order to provide lighting for homes, power for the village school, the village health centre and for agricultural machinery. Both projects would have used renewable energy to contribute towards the alleviation of poverty in the villages concerned, while also contributing to sustainable development.

The bureaucracy behind CDM projects includes certification by consultants, acceptance by governments and approval by a United Nations secretariat, all of which push up the cost of a project. For small projects involving less than $250,000 worth of credits, fees for deal makers, consultants and lawyers can be far higher than the cost of installing equipment to clean up emissions.[5] Sure enough, in the two Philippines case studies, not enough emission credits would have been generated to justify the high costs required to make them into CDM projects. The researchers' conclusion was that: 'These two case studies suggest that the CDM may not provide the carbon revenue which will bolster significantly the implementation of poverty alleviation projects that are also climate-friendly.'[6]

While renewable energy projects such as those in the Philippines case studies may be the most socially beneficial and environmentally sustainable type of energy systems, in the short term they involve greater investment and infrastructure development, and their returns are not guaranteed. This leads to a conflict within the CDM between affordability and sustainability: a conflict that sustainability cannot win. This is because, as Ben Pearson of the non-governmental organisation CDM Watch said: 'despite the rhetorical trimmings the CDM is a market, not a development fund nor a renewables promotion mechanism. Its aim is to provide tradable emission reduction credits at the lowest cost in a limited timeframe.'[7]

## A lucrative harvest?

The most popular CDM projects to date have been hydroflorocarbon (HFC-23) capture projects. About $500 million of the emission credits delivered by the World Bank in 2006 came from two very large projects for HFC-23 destruction in China.[8] These stand the greatest chance of guaranteeing financial returns; the equipment can be bolted on relatively cheaply to existing systems, and they deliver a high amount of credits, as these credits are counted in carbon dioxide-equivalence and HFC-23 is nearly 12,000 times as potent a greenhouse gas as carbon dioxide.

Capturing and destroying such a potent greenhouse gas as HFC-23 is, of course, a good thing. However, is it the best use of resources intended to create more sustainable energy systems? In an article in *Nature* Michael Wara, an associate at San Francisco-based legal firm Holland & Knight, said he

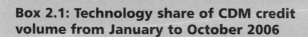

## Box 2.1: Technology share of CDM credit volume from January to October 2006

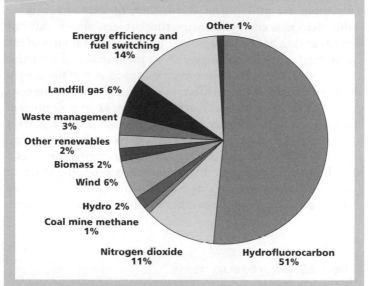

- Other 1%
- Energy efficiency and fuel switching 14%
- Landfill gas 6%
- Waste management 3%
- Other renewables 2%
- Biomass 2%
- Wind 6%
- Hydro 2%
- Coal mine methane 1%
- Nitrogen dioxide 11%
- Hydrofluorocarbon 51%

Source: World Bank, *State and Trends of the Carbon Market 2006 (Update)*, October 2006, p. 12.

thought that, for each dollar spent, HFC-23 capture projects are not very effective at reducing emissions.[9] Wara proposed an alternative approach to cutting HFC-23 emissions – paying producers directly for the cost of installing the technology to capture and destroy the gas. He said: 'This technological solution would cost the developed world less than €100 million, saving an estimated €4.6 billion in CDM credits that could be spent on other climate-protecting uses.'[10]

Wara points out that the global carbon market, which is the overall framework for the CDM, was supposed to create incentives to invest in infrastructure for low-carbon energy in developing countries. The dominance of HFC-23 capture rather than new energy infrastructure projects shows that this is not working. HFC-23 capture is likely to make up less of the CDM market in the coming years as the number of available projects dwindles. To ensure that its space is not filled by a similarly quick-fix but ineffective technology, Wara suggests that, if the global carbon market is to make any significant contribution to reducing emissions, it has to be made a market for carbon dioxide only. Carbon dioxide is particularly harmful to the climate as it is emitted in large quantities and has a long lifetime.[11] Whether this would then push the price of carbon high enough to make small-scale projects (such as those tried in the Philippines) more viable remains to be seen.

## Box 2.2: Short-term gains

Capturing the greenhouse gas methane from landfill sites has been popular with CDM investors as it does not involve significant new finance and projects can be set up quite quickly. One controversial example of landfill gas capture has been the Durban Solid Waste project in South Africa. Opponents argue that, far from being a welcome environmental adaptation, the project is an unwanted health nuisance.

The Durban Solid Waste project involves capturing methane from three landfill sites, then using the methane in electricity generators, cutting back on emissions from the coal that these generators would have used otherwise.

One of the landfill sites is in the middle of a residential area, with a school just across the street. After opposition to the site, city authorities had given assurances that it would close; yet when the time came, a permit to extend its life was granted.

The site was licensed to receive only domestic waste. However, it has been reported that medical waste, sewage sludge, private corporate waste and large shipments of rotten eggs have also ended up there. Cadmium and lead emissions at the site are reportedly over legal limits, and concentrations of methane, hydrogen chloride, and other organic and inorganic compounds including benzene and toluene, trichloroethylene and formaldehyde are high.

Even though methane is captured the landfill site still emits other pollutants, including carbon monoxide and various hydrocarbons. The site's neighbours have reported many health problems. Local resident Sajida Khan said:

> To gain the emissions reductions credits they will keep this site open as long as possible. Which means the abuse will continue as long as possible so they can continue getting those emissions reductions credits. To them how much money they can get out of this is more important than what effect it has on our lives.

Source: Larry Lohmann, 'Marketing and Making Carbon Dumps: Commodification, Calculation and Counterfactuals in Climate Change Mitigation,' *Science as Culture*, Vol. 14, September 2005, pp. 222–4.

## Box 2.3: Problems with the overall framework

In India, CDM projects have been proposed that generate electricity from biomass such as rice husk, cotton sticks, chilli waste, mustard sticks and wood of the plant *Prosopis juliflora*. Project design documents say that the projects will use 'surplus biomass', which is 'otherwise under-utilised or burnt with no commercial value'. But the science and environment magazine *Down to Earth* has pointed out that there is no such thing as 'surplus biomass'. All potential raw materials planned for biomass gasification have competing uses – rice straw is used as fodder; mustard sticks and *Prosopis juliflora* are the fuelwood of the very poor.

*Down To Earth* journalists visited a factory in Karnataka that uses biomass and was hoping to apply for CDM credits. They found that residents in a nearby village were accusing the factory of large-scale deforestation. One villager told them:

> First, the plant cut the trees of our area and now they are destroying the forests of ... other places. They pay Rs 550 per tonne of wood, which they source using contractors. The contractors, in turn, source wood from all over the state.

Another said: 'Now poor people find it difficult to get wood for cooking and other purposes.'

Factory staff deny these allegations. Plant manager Amit Gupta said: 'if there is deforestation, then local people are to be blamed because they are supplying the wood to us.'

*Down to Earth* concluded that:

> Power generation through decentralised projects using biomass energy, as against dirty fossil fuels, is definitely a way to the future. But for this, the project design needs a supportive policy framework which ensures biomass is sustainably harvested; that the benefits of raw material supply and even power generation go to local communities.

Source: *Down to Earth*, 15 November 2005.

## Rush to the bottom

In countries that play host to projects, the CDM may actually undermine the likelihood of policies that encourage sustainable development. Credits are only earned for projects that reduce emissions below what they would have been if the CDM project had not taken place. This means that even if a country was interested in developing policies geared towards reducing emissions, the CDM may make it think again. For example, if a government introduced tougher vehicle emission standards for public transport it could not then go on to apply for CDM credits for the sector, as emission standards would be seen as 'business-as-usual' rather than a project that had come about as a result of the CDM.[12] This may lead to business-as-usual scenarios that are as high emitting as possible, so that proposed CDM projects will make the biggest emission cuts possible, thus reaping more credits. In a 2005 article in *Down to Earth* magazine, Ritu Gupta, Shams Kazi and Julian Cheatle noted: 'The current design [of the CDM] provides countries with perverse incentives to keep polluting as long as they have the money to pick up carbon credits. In this respect, is not CDM actually against sustainable development?'[13]

Also, countries that are interested in hosting CDM projects are allowed to set their own criteria for sustainable development and to judge for themselves whether a project meets these. Host countries should, of course, participate in the overall shaping of a mechanism of which they are a fundamental part. However, the competition for this new income stream can be another incentive to loosen up sustainable development criteria, particularly for countries short of money or with debt repayments to meet. In a paper published by the International Institute for Environment and Development, Jessica Ayres et al. summed this up as:

> Governments face the dilemma of setting demanding sustainable development criteria and running the risk of losing investments to other developing countries with less demanding standards, or setting less stringent standards and thus generating little benefit at the local level.[14]

## A new source of development funding?

One of the justifications for the CDM is that it brings much-needed resources to poorer countries. A look at the location of CDM projects, however, suggests that it may not be the most effective vehicle for doing this. To date, most CDM projects have taken place in middle-income countries, particularly those undergoing rapid industrialisation. The HFC-23 capture projects – the big credit generators – are concentrated at a handful of sites, mainly in Asia (in China, which had a 60 per cent share of the CDM market in 2006, in India, and in South

Korea) and in Latin America (in Mexico and Brazil).[15] Yet the continent where energy systems and sustainable development are most badly needed, and where climate change hits the hardest, is Africa. Even when fewer HFC-23 capture projects become available, as is expected to be the case over the next few years, the next big CDM investment category may be carbon capture and storage projects which, again, are likely to be located outside Africa. As of the end of October 2006, there were 19 projects from sub-Saharan Africa being considered as possible CDM projects, out of a total of 1274 projects under consideration for all developing countries.[16] The World Bank has said that:

> Most developing countries can only deliver small CDM-eligible projects. The high transaction costs and high risks involved in delivering carbon from these projects means that most of the smaller and poorer of the Bank's client countries will be unable to benefit from carbon finance as a catalyst for investment in clean technologies.[17]

The CDM does have something specifically targeted at the poor: 2 per cent of the income from each deal will go into an Adaptation Fund to be spent on poorer countries that have ratified the Kyoto Protocol. As an international tax this may be setting a valuable precedent, but for now its contribution to dealing with climate change is minimal.

## Defining the agenda

Even if the CDM does not currently contribute large funds to climate change adaptation, it still influences the path of development taken by countries that host projects. Firstly, as outlined above, countries may reduce their sustainable development

## Box 2.4: Registered CDM projects by country, February 2007

| Country | Number of projects | Country | Number of projects |
|---|---|---|---|
| Argentina | 6 | Jamaica | 1 |
| Armenia | 2 | Malaysia | 12 |
| Bangladesh | 2 | Mexico | 73 |
| Bhutan | 1 | Mongolia | 1 |
| Bolivia | 1 | Morocco | 3 |
| Brazil | 88 | Nepal | 2 |
| Cambodia | 1 | Nicaragua | 2 |
| Chile | 14 | Nigeria | 1 |
| China | 37 | Pakistan | 1 |
| Colombia | 6 | Panama | 4 |
| Costa Rica | 2 | Papua | |
| Cyprus | 2 | New Guinea | 1 |
| Dominican | | Peru | 3 |
| Republic | 1 | Philippines | 8 |
| Ecuador | 8 | Republic | |
| Egypt | 2 | of Korea | 10 |
| El Salvador | 2 | Republic of | |
| Fiji | 1 | Moldova | 3 |
| Guatemala | 5 | South Africa | 6 |
| Honduras | 10 | Sri Lanka | 4 |
| India | 162 | Tunisia | 2 |
| Indonesia | 8 | Uganda | 1 |
| Israel | 3 | Viet Nam | 2 |

Source: United Nations Press Release, 12 February 2007.

practice and criteria in the competition to secure CDM projects. Secondly, the very presence of a project defines choices for energy generation or other behaviour. Considerable assumptions – and thus narrowing of possibilities – are made over future development paths. For example, if a country does actually get a CDM project, the amount of emissions credits that the project earns is based on a prediction of what might have been if the project had not taken place. This could assume that a country would have continued to rely on fossil fuels. It could assume that any social movement campaigning for a low carbon economy would fail. The without-project scenario is presented, says Larry Lohmann of the non-governmental organisation The Corner House, as 'singular, determinate and a matter for economic and technical prediction'.[18] This, says Lohmann, 'hampers thinking about broader social and industrial change' for the future is, effectively, already decided.[19]

### Box 2.5: Carbon colonialism

In Brazil, a company called Plantar applied for emission credits for using charcoal rather than coal as an energy source for making pig iron. The charcoal was made from eucalyptus trees grown on a plantation owned by Plantar.

Eucalyptus plantations can be ecologically destructive, causing loss of biodiversity, contaminating local environmental resources, and draining and polluting local water sources. They can also be socially destructive, as they enclose land previously used by local people.

In 2003 a coalition of non-governmental organisations, unions, church groups and individuals wrote to the World Bank, which was promoting the Plantar project, saying:

> Corporations like Plantar S.A. installed themselves in our states in the 1960s and 1970s during the military dictatorship, taking advantage of attractive tax incentives. Unfortunately, local communities who were directly impacted by the actions of the corporations were never consulted about

whether they wanted this type of project for their region. The result was that [indigenous peoples] were expelled from their lands ... increasing unemployment and, consequently, the despair of these populations who lost their lands and were left without their biodiversity and without their water, on which they were dependent.[1]

Such practices, where 'land is commandeered in the South for large-scale monoculture plantations which act as an occupying force in impoverished rural communities dependent on these lands for survival', have been dubbed 'carbon colonialism' by their critics.[2]

Emissions credits are a new globally traded commodity. Clear property rights are needed in order to trade this commodity – you must own the eucalyptus plantation, for example, in order to claim credits for its role as a carbon sink. Yet traditional management of resources has not necessarily involved such formal boundaries. In 2004, land that had been 'rented' to a plantation company by the state government in Minas Gerais in Brazil was occupied by people who lived in the area. People had been expelled when the area was turned into a plantation, leaving them with no access to land that had provided many of their basic needs but had never been registered as their property.

On the other side of the world, plantation owners in New Zealand challenged their government in 2003 over who owned the carbon in 200,000 hectares of trees that were eligible to earn emissions credits. The plantation owners claimed the government was trying to steal NZ$2.6 billion from them with a stroke of the pen. This was, they said, 'possibly the largest private property theft in New Zealand's history'.[3]

Sources:

1. Open letter to those responsible for, and investing in, the Prototype Carbon Fund (PCF), quoted in CDM Watch, *The World Bank and the Carbon Market: Rhetoric and Reality*, April 2005, p. 14.
2. Heidi Bachram, 'Climate Fraud and Carbon Colonialism: The New Trade in Greenhouse Gases', *Capitalism Nature Socialism*, December 2004, p. 7.
3. *Business Today (New Zealand)*, 30 December 2003 quoted in Larry Lohmann, 'Marketing and Making Carbon Dumps: Commodification, Calculation and Counterfactuals in Climate Change Mitigation', *Science as Culture*, Vol. 14, September 2005, p. 214.

## Sticking plasters on a gaping wound

The non-governmental organisation the WWF believes that it is still possible to make the CDM work, stating that: 'If designed correctly, CDM ... projects can play a valuable role in promoting the spread of sustainable energy technologies both within and outside the industrialised world.'[20] Together with other non-governmental organisations, the WWF has launched the Gold Standard – a code of practice that uses environmental and social indicators to check that a CDM project contributes to sustainable development. One of the most high profile, and at first glance successful, Gold Standard projects is the project in Kuyasa, South Africa that was referred to in the Introduction, where Muzelli and his neighbours were on the receiving end of funding for low-emission projects. According to Graham

Erion, who has carried out extensive research on the South African carbon market, this was the only African CDM project he was aware of that was actually supported by the local residents. The project incorporated extensive community consultation, it created new jobs, and it led to specific improvements for the households involved. 'On the face of it,' said Erion, 'there is very little not to like about [it].'[21] Yet this good beginning may be the place where any promise in the project ends.

Erion found that Kuyasa's future was far from certain, as its funding as a carbon project was not sustainable. Much of the budget for the project's first phase came from sources other than emission credits – sources that would not necessarily be available in the future. These included government departments and the social responsibility budget of a multinational company. The project's second phase was likely to receive even less funding from emission credits. This was because it would not be able to compete for CDM support with the likes of South Africa's landfill projects, which are profitable at a lower carbon price as methane from landfill generates so many emission credits. Erion said:

> although [Kuyasa] is exactly the type of project that many people hoped the CDM could deliver, now that it exists the carbon market simply cannot support it. The basic problem is that the projects that are out there are driven first and foremost by economic considerations and thus are driving down the price of carbon.[22]

He continued:

> The Gold Standard seems to have become the victim of the very scourge it was set up to avoid: the propensity of Northern governments to only invest in projects that

offer maximum return on investment with little added environmental and social benefits. Worse, it has now given these projects greater legitimacy and demand.[23]

## Box 2.6: A tarnished standard?

In December 2000 the city of Cape Town released an Integrated Waste Management Plan. This suggested that if biodegradable organic material was separated from non-organic material, the city could vastly decrease its need for landfill space while also capturing a much higher amount of methane. Methane is generated from rotting organic material but when this is mixed in with non-organic material, as is typical practice in landfills, the amount that can be captured is reduced. Trying to capture methane from a regular landfill is, according to Walter Loots, head of Solid Waste for Cape Town and lead author of the Integrated Waste Management Plan, 'an inefficient solution to an avoidable problem'.

Despite this conclusion, South Africa's Belville South landfill gas capture project has been awarded the Gold Standard as a 'renewable energy' project that makes a 'significant contribution in terms of the local, regional and global environment'.

One reason for this apparent contradiction – awarding the Gold Standard to a project that runs contrary to its host city's aspirations for waste management – is that the city of Cape Town does not have the resources to institute a large-scale waste separation scheme. Should the Gold Standard be awarded to a project that is not the most sustainable solution but is the only one that can be afforded?

Graham Erion's report *Low Hanging Fruit Always Rots First: Observations from South Africa's Crony Carbon Market* says that this example illustrates a failure of imagination in the carbon market: a failure to produce forward-thinking projects with long-lasting social and environmental benefits for the community. Erion says:

> A CDM project that provided the capital for a municipality to put in a widespread recycling and waste separation system would have undeniable environmental and social benefits. The space required for landfills would be vastly reduced and without the organic material rotting they would cause much less nuisance to surrounding areas. In addition to improving productive methane capture from the sorted organic material, the better solution for avoiding climate change, the very act of sorting would create thousands of employment opportunities.

Source: Graham Erion, *Low Hanging Fruit Always Rots First: Observations from South Africa's Crony Carbon Market*, Center for Civil Society, University of KwaZulu-Natal, South Africa, 2005, pp. 39–41.

The WWF is a credible and experienced organisation, and is fully aware that the CDM is a market-based mechanism. A senior policy officer for the WWF's Climate Change Programme said:

> We know that [the] carbon trade is going to be a multi-billion dollar industry. ...The Gold Standard can help

ensure that some of that money – perhaps even most – goes to projects with high environmental and developmental value, which greatly enhance energy security.[24]

However, rather than the direction of money being influenced by the environmental and social parameters of the Gold Standard, it seems like the opposite is happening – the Gold Standard is being compromised in order to stay in the market. Further pressures on the Gold Standard will come from trading rules beyond the carbon market, where trends are towards less intervention, not more, in the form of labelling or environmental standards.[25] It may be difficult to make the more regulated Gold Standard projects commercially viable or even possible.

## Box 2.7: The carbon market

The carbon market is part of the wider strategy of pricing environmental resources. For example, the United Nations has promoted the idea of 'natural capital'[1] as a way of valuing environmental goods so that they can be included in economists' equations. Klaus Töpfer, former executive director of the UN Environment Programme, said: 'The goods and services delivered by nature, including the atmosphere, forests, rivers, wetlands, mangroves and coral reefs, are worth trillions of dollars. When we damage natural capital, we not only undermine our life support systems but the economic basis for current and future generations.'[2] Richard Caines, a manager at the International Finance Corp. agrees. He describes biodiversity as a 'genuine business opportunity'.[3]

While the damage caused to the Earth's climate systems as a result of burning fossil fuels clearly demonstrates the need to value nature more, an economic assessment is not necessarily the best way to do this. Setting a price tag means only valuing something to the extent that it provides commodities for human beings – for example clean water, food, areas for recreation. In nature, ecosystems or species of no obvious use to human beings are therefore considered to be of no value, despite the fact that they have value in their own right, regardless of their worth to human beings. Similarly, market-based decision making means that those who make little obvious contribution to an economic system – the poor or future generations – have no value. When international climate strategies are dominated by market mechanisms, they become, says the non-governmental organisation Carbon Trade Watch, a 'numbers game' where 'those disempowered by the lexicon of free-market environmentalism and cost-benefit analysis end up short'.[4]

Sources
1. *Environment News Service*, 25 September 2006.
2. *Environment News Service*, 20 June 2005.
3. *Reuters*, 27 March 2006.
4. Carbon Trade Watch, *Hoodwinked in the Hothouse: the G8, Climate Change and Free-market Environmentalism*, Transnational Institute briefing series No 2005/3, June 2005, p. 37.

## Our hero's tragic flaw

Trading in emissions credits through schemes such as the Clean Development Mechanism will become increasingly popular as richer countries use up easier methods of cutting emissions and rely instead on paying for change to happen elsewhere. In December 2005 *BusinessWeek Online* predicted that: 'as 2012 approaches and companies in the West realise it's cheaper to buy credits than to clean up at home, purchases of credits from developing countries are expected to soar.'[26]

Yet possibly the most tragic flaw of the CDM is that the accounting system behind it may not actually add up. It entails an extremely vague process of working out the number of credits that are earned by measuring predicted and possible emissions levels and judging 'what might have been' in terms of emissions as part of some possible future scenario. This is incredibly difficult to assess and involves a huge range of variables. Larry Lohmann pointed out that:

> Physical actions (for instance, planting biomass for power plants) bring about social effects (for example, resistance among local farmers, diminished interest in energy efficiency among investors or consumers, loss of local power or knowledge), which in turn bring about further physical effects (for instance, migration to cities, increased use of fossil fuels) with carbon or climatic implications. ... No basis exists in either physical or social science for deriving numbers for the effects on carbon stocks and flows of such social actions.[27]

Emissions credits can easily be created by, for example, re-thinking ideas for what might have happened without a CDM project. Similarly, claims that a project would not have happened without the CDM (which is a requirement if it is to claim credits) are also easy to make. A report by Lohmann gave one such example where consultants in a Latin American country covered over the name of a hydroelectric dam in a copy of a national development plan.[28] This was an attempt to show that the dam was not already planned or 'business as usual' and therefore was deserving of carbon credits.

Lohmann's concerns are shared by an unlikely source – the CDM's Executive Board (EB) itself, which has introduced regular checks on the organisations that verify CDM projects. These are known as designated operational entities, or DOEs. The Board performed spot checks on at least two DOEs after finding what it termed 'non conformities' in operational structures and in the ability to validate offset projects and verify whether actual emission reductions had taken place.[29] The work of the DOEs seems quite shambolic: in the case of the Plantar project discussed in Box 2.5, the scheme was recommended to the CDM Board even though the company that verified it admitted an inability to arrive at a final conclusion about the project's climate benefit.[30] One source in the carbon market, who spoke to the carbon analysis service Point Carbon on condition of anonymity, said: 'The problem is, the EB doesn't trust the DOEs. They put the rules in place and when a DOE follows the validation procedure the EB doesn't believe them.'[31] The CDM's Executive Board has said that it will ensure 'on-site assessments at least once every three years',[32] but will it have the resources to do this? And, again, is this really the best way to direct resources for stabilising the climate?

While the CDM is a key part of the carbon market, it alone cannot be held to blame for wider failings, such as the difficulty of verifying emissions reductions. It does, however, illustrate that one of the key mechanisms for reducing emissions is severely flawed. It should be ringing alarm bells over whether trading mechanisms can deliver the level of emissions reductions needed in the timescale that is available. For, while we're waiting to learn more lessons from more trading schemes, the possibility of a stable climate may be slipping further out of reach.

# 3
# The World Bank

We have an opportunity today, to think outside the box and find new ways, practical solutions, to promote the generation and diffusion of low carbon technologies and the integration of climate concerns in development strategies. Let's work together for a climate friendly future.

Paul D. Wolfowitz, former World Bank President[1]

## Why focus on the World Bank?

While the CDM is but one element in the carbon market, the institution that provides the overall market framework is the World Bank. The World Bank is hugely powerful. It attaches conditions to its loans, which define policy and land-use choices for its debtors. It funds projects and infrastructure that last for many decades. These policies, land-use choices, projects and infrastructure influence a country's capacity to adapt to climate change, and influence levels of greenhouse gas emissions. The World Bank also sets the trend for decisions made by aid donors and private banks and funders, as its involvement shows others that a country is a reliable place to invest.

In 2005, the Bank's influence was formally extended to the climate crisis when the world's most industrialised countries, the G8, asked it to develop a new framework for addressing the problem. As a global institution wielding great power the Bank may seem an obvious candidate for such a role. But there is a problem with this. The Bank actively promotes the use of coal, oil and gas – the fossil fuels that are having such a disastrous impact on the climate. Can it really address climate change, then, when its own practices are such a large part of the problem?

## Funding fossil fuels

The World Bank has played a significant part in funding coal, oil and gas-based energy systems. Between 1994 and 2003 the World Bank Group approved over $24.8 billion in financing for fossil fuel extractive and power projects.[2] During this time it approved just $1.06 billion in renewable energy projects.[3] And, if redirected to small-scale solar installations in sub-Saharan Africa, just one year's worth of World Bank spending between 1992 and 2003 could have provided 10 million people on the continent with electricity.[4] When former World Bank President James D. Wolfensohn said we have 'to remain realistic: renewable energy is expensive,'[5] he seemed to forget the fact that his former institution's choice of energy subsidies helps make it so.

By 2004 fossil fuel projects still made up 94 per cent of the Bank's energy portfolio,[6] even though by that stage the UN's Intergovernmental Panel on Climate Change was saying that the greenhouse gases emitted when fossil fuels are burned were changing the atmosphere in ways that would affect the climate. The Bank's energy portfolio included spending on infrastructure planned to last a number of decades: longer than the window of the next 10–15 years, the time in which we have to dramatically reduce emissions from these very same fossil fuels.

## The Bank's own paradox

In 2000 the Bank launched an Extractive Industries Review (EIR) to investigate whether its investments in oil, gas and mining – the investments that are so detrimental to the climate –

were consistent with its overall objective, which is to achieve poverty alleviation through sustainable development. The EIR was headed up by Dr Emil Salim, a former Minister of the Environment for Indonesia. Workshops were held in a range of locations around the world, including Brazil, Hungary, Mozambique, Indonesia and Morocco. Research projects were commissioned, project sites were visited, and consultations with stakeholders were held. Although this review was not specifically focused on climate change, any recommendations over the extractive industries would have profound implications for the Bank's energy policy, and thus for greenhouse gas emissions.

When the EIR was published in 2004 its conclusions were striking. It found that large energy infrastructure projects of the kind favoured by the Bank had caused substantial damage at local levels, including environmental degradation, social disruption, and conflict.[7] The EIR suggested that the Bank 'must take certain steps to rebalance its institutional priorities'.[8]

Perhaps the most dramatic conclusion from the Review was that:

> The WBG [World Bank Group] should phase out investments in oil production by 2008 and devote its scarce resources to investments in renewable energy resource development, emissions-reducing projects, clean energy technology, energy efficiency and conservation, and other efforts that de-link energy use from greenhouse gas emissions. During this phasing out period, WBG investments in oil should be exceptional, limited only to poor countries with few alternatives. And the WBG has for the last few years not invested in new coal mining development. This should continue.[9]

This recommendation should have carried great weight, coming as it did in a review headed by an eminent statesman and commissioned by the Bank itself. It seemed as if those who understood the Bank considered that a different direction was not only necessary but, most importantly, was possible. Unfortunately the Bank's management did not see it this way. Rather than end involvement in extractive industries they decided that more of the same was remedy enough: 'by staying engaged in oil and coal we can have an influential role.'[10] Given that the Bank's own review process had raised serious questions over its role so far, this response was disappointing to say the least.

## Moving targets

The Bank did seem to accept one suggestion from the Review. This was the suggestion that it should increase investments in renewable energy by 20 per cent each year. At a 2004 International Conference on Renewable Energies, the Bank said it would commit to an average growth rate of 20 per cent per year over the following five years in its annual financial commitments for renewable energy and energy efficiency projects. Peter Woicke, managing director of the World Bank Group said: 'Our strategy – through programs and policies – will aim to ensure that renewable energy and energy efficiency are seen as ... essential ingredients in the energy choices of our member nations, not marginal considerations.'[11] Although this sounded like a move towards the cleaner energy sources that are needed if greenhouse gases are to be reduced, the figures have not added up since. In a report the following year, Friends of the Earth US found that the Bank still had a long way to go if it was to demonstrate a real commitment to cleaner energy.

Friends of the Earth began by looking at how the World Bank constructed its figures. Within the World Bank Group are five distinct agencies: the International Bank for Reconstruction and Development (IBRD), the International Development Agency (IDA), the International Finance Corporation (IFC), the Multilateral Investment Guarantee Agency (MIGA), and the International Centre for Settlement of Investment Disputes (ICSID). The term 'World Bank' usually only refers to the IBRD and IDA, while the term the 'World Bank Group' includes all five of these institutions.

Friends of the Earth found that renewable energy and energy efficiency funding from the IBRD and IDA for 2005 was $223 million: only a 7 per cent increase over the baseline of $209 million – and far from the promised 20 per cent.[12] Even this figure was inflated as it included financing from other sources which the World Bank implemented or co-implemented. While the Bank should lever in funding for renewables from other sources, it is slightly misleading to describe this as World Bank funding. Such a claim also obscures the balance between fossil fuels and renewables in the World Bank's own energy spending. According to Friends of the Earth, support for renewable and energy efficiency projects through the World Bank's own funds – IBRD and IDA financial resources alone – actually totalled only $109 million. And $87 million of this was for just one project.[13]

Meanwhile, financing from the World Bank Group's International Finance Corporation (IFC) and Multilateral Investment Guarantee Agency (MIGA) was not included in its target for investment, and only 2 per cent of IFC energy funding for 2005 went to renewable energy or energy efficiency.[14] Taking the World Bank Group as a whole, Friends of the Earth found that renewables and energy efficiency made up only 9 per cent of overall energy financing for 2005.[15] Even this dismal level of funding, which fell far short of the Bank's own target, did not seem to be aimed at those most in need. Sub-Saharan Africa, which has some of the highest poverty and

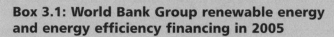

## Box 3.1: World Bank Group renewable energy and energy efficiency financing in 2005

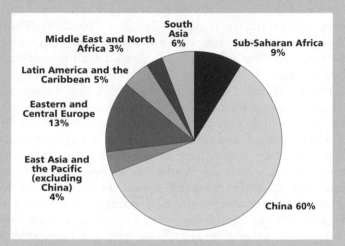

Source: Elizabeth Bast and David Waskow, *Power Failure: How the World Bank is Failing to Adequately Finance Renewable Energy for Development*, Friends of the Earth – United States, October 2005, p. 14.

lowest levels of electrification in the world, received only $22 million of the IBRD and IDA's funding for renewable and energy efficiency projects.[16]

## Fossil fuels and export

So why is the Bank so apparently reluctant to invest in the renewable energy and energy efficiency projects that, as Chapter 5 outlines, would set poorer countries on a road to increased security, and might even achieve the Bank's overall objective of poverty alleviation? After all, the Bank itself has acknowledged the importance of renewable energy. Katherine Sierra, one of the Bank's vice presidents has said:

> sources of energy must be secure if [developing countries'] economies are to develop and living standards are to improve. ... Renewable energy can enhance a country's energy security by reducing energy import requirements; reduce supply risks by diversifying its energy portfolio and protecting precious financial resources.[17]

A look at the energy priorities of the Bank's key decision makers reveals a possible answer to this question. Although the World Bank Group is owned by more than 180 member governments, the number of voting shares held by each country is determined by its gross national product and its financial contributions to the Bank. Therefore, those countries that are richest – generally the most industrialised and the most dependent on fossil fuels – wield influence at the Bank. This may explain why the World Bank used public money, in the name of development aid, to fund the Baku–Tblisi–Ceyhan (BTC) pipeline, which carries oil from the Caspian sea through Azerbaijan, Georgia and Turkey with arguably few local benefits.[18] The pipeline was heavily promoted by the US administration, which is keen to secure sources of non-Middle Eastern oil. A report by several non-governmental organisations said the pipeline will ensure that: 'Caspian oil is burnt in the car and truck engines and power plants of Western Europe, the United States and the Far East.'[19] The report estimated that when the pipeline is operating at full capacity the one million barrels of oil that it carries every day will contribute 160 million tons of carbon dioxide to the atmosphere every year.

Moreover, the Bank has to ensure the repayment of debts owed both to itself and to the corporations based in the countries that are its major shareholders. Oil and gas projects can earn hard currency for this debt repayment. The Sustainable Energy & Economy Network (SEEN) found that 82 per cent of all oil extractive projects funded by the World Bank Group between 1992 and 2004 were export-oriented (see Box 3.2).

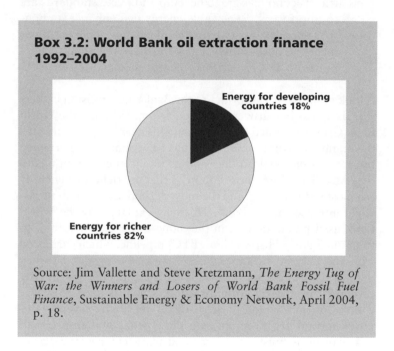

**Box 3.2: World Bank oil extraction finance 1992–2004**

Energy for developing countries 18%

Energy for richer countries 82%

Source: Jim Vallette and Steve Kretzmann, *The Energy Tug of War: the Winners and Losers of World Bank Fossil Fuel Finance*, Sustainable Energy & Economy Network, April 2004, p. 18.

A SEEN report on World Bank fossil fuel finance described how Bangladesh owed over $74 million to energy companies Halliburton, Shell and Cairn Energy, and faced tremendous pressure from the World Bank to develop its gas reserves for export as 'a major potential source of foreign exchange earnings'.[20] The World Bank has said it sees energy as crucial for poorer countries' development, yet only 17 per cent of Bangladesh had access to electricity at the time.[21] Surely then, if Bangladesh's gas was going to be used anywhere, it should be used at home?

A further sad irony behind developing and exporting fossil fuels is also clearly lost on the Bank. For the resource that Bangladesh was encouraged to develop in order to pay back debt will accelerate the climate change that may destroy low-lying Bangladesh and undermine its ability to pay back debts in the longer term anyway. Continued support for fossil fuels is, as much as anything, bad economics.

## Box 3.3: An uncomfortable challenge?

Global energy markets based on fossil fuels form an integral part of the infrastructure of globalisation. Poor countries get hooked into this infrastructure through their reliance on oil, coal and gas imports, and end up in a nexus of dependency relationships with other nations, multilateral donors and foreign companies. Financing fossil fuel projects can account for up to 40 per cent of total bilateral debts to export credit agencies; meaning that developing country governments become tied to policies dictated by the multilateral development banks and the major OECD donor governments. Imagine a poor country that suddenly shifted all energy investments into locally based renewable energy systems. Before long, the economic perspective and priorities of that country would shift drastically. No need to borrow hundreds of millions to develop expensive energy plant and infrastructure systems; massive national energy bill savings as consumption of imported oil and gas plummets; greatly reduced geopolitical vulnerability to international events such as OPEC price hikes, or war, that could affect energy supplies and economic stability. In short; a massive tether to the global economy would be severed.

This scenario could lead to a more systemic rethink of the development process per se. Less debt and spending on energy imports would reduce the need to generate foreign exchange revenue through exports. Local economies could focus more on meeting local and domestic needs, without needing to look to foreign investors or markets to pay for their development needs. If such a scenario emerged in one country it would pose an uncomfortable challenge to established global thinking about development. If it emerged in several countries, it would begin to have major effects on global energy markets, as well as international imports and exports of goods and services. Were it to spread across the developing world, it would fundamentally transform economic globalisation.'

Source: Andrew Simms, Julian Oram and Petra Kjell, *The Price of Power: Poverty, Climate Change, the Coming Energy Crisis and the Renewable Revolution*, nef (the new economics foundation, London, 2004, pp. 23–4. www.neweconomics.org.

## Carbon contradictions

Despite its unhealthy interest in fossil fuels, the World Bank has expressed great interest in taking up the G8's suggestion and playing a leading role in tackling climate change. The main mechanism through which it does this is the carbon market, where the Bank is one of the key players. This fits in with the Bank's traditional approach for development, which is to encourage countries to get money by developing assets as goods for export, as in the case of Bangladesh's gas. This time the asset at stake is emissions credits – the right to emit greenhouse gases. The Bank has said: 'While many developing countries have become actively engaged in carbon finance, others have barely recognised the opportunity to earn carbon emissions reduction credits and exploit them as export "commodities".'[22]

The Bank itself is a massive buyer of carbon credits, and its carbon funds form a large part of the CDM market described in Chapter 2. The Bank has also administered country carbon funds for the Dutch, Italian, Spanish and the Danish Governments. From July 2004 to July 2005 alone, carbon funds administered by the Bank grew from $413.6 million to about $914.7 million.[23]

The Bank has three objectives for its carbon finance activity: to ensure that it contributes to sustainable development through, for example, getting low-carbon technologies to developing countries; to expand the carbon market; and to strengthen the capacity of developing countries to benefit from the market. The Bank has said that: 'carbon finance initiatives are part of a larger global effort to combat climate change, and go hand in hand with the Bank's mission to reduce poverty and improve living standards in the developing world.'[24] As Chapter 2 illustrated, claims that the carbon market will contribute to sustainable development or deliver significant new finance into the hands of the poor are problematic.

The World Bank's – and the world's – first carbon fund was the Prototype Carbon Fund (PCF). This was established by the Bank in 1999: 'to show how a market for carbon emission credits for developing countries can work under the Kyoto Protocol's proposed flexible mechanisms'.[25] Credits from the PCF can be used towards Kyoto emission targets, while businesses can use them in markets such as the European Union Emissions Trading Scheme. Unfortunately, in so far as it shows anything, the PCF shows that the market approach is seriously flawed (the CDM projects featured in Box 2.2 and Box 2.5 were both PCF projects).

Research carried out by the non-governmental organisation CDM Watch suggested that emissions reductions from the PCF pale into insignificance when compared to the likely emissions from fossil fuel projects funded by the World Bank. For example, in the same year that the PCF was launched the Bank approved over $551 million for the controversial Chad–Cameroon oil pipeline.[26] CDM Watch said that:

> The financing package for Chad–Cameroon is about three times the capitalisation of the PCF and its expected lifetime emissions of approximately 446 million tonnes of carbon dioxide are roughly three times the 142 million tonnes that will allegedly be reduced by PCF projects in total.[27]

Many investors in the PCF also receive far greater amounts back from the World Bank for fossil fuel projects (see Table 3.1). CDM Watch said that many of these projects 'will help lock those countries into a fossil fuelled energy path and lead to greenhouse gas emissions orders of magnitude greater than the PCF projects claim to be reducing'.[28] CDM Watch also pointed out that 'the PCF investors get a carbon credit for the PCF projects, but no debit for those involving fossil fuel extraction and use'.[29]

The PCF also offers carbon credits through the CDM for projects involving large dams. The World Bank considers these to be a clean energy option, despite studies showing that large hydropower projects are a net greenhouse gas emitter, particularly in tropical climates, where the rapid breakdown of plant matter in reservoirs creates large amounts of methane.[30] The Bank put $449 million towards

## Table 3.1 CDM compared with World Bank fossil fuel funding

| Corporation | PCF contribution (US$ million) for CDM and Joint Implementation* projects 1999–2004 | Received from WB for fossil fuel projects 1992–2002 (US$ million) |
| --- | --- | --- |
| Mitsui (PCF and BCF) | 16 | 1,807.5 |
| BP | 5 | 938.8 |
| Mitsubishi | 5 | 403.6 |
| Deutsche Bank | 5 | 165.6 |
| Gaz de France | 5 | 138.9 |
| RWE | 5 | 138.9 |
| Statoil | 5 | 242.3 |
| Total | 46 | 3,834.6 |

*Joint Implementation (JI) projects work in a similar way to CDM projects except that they take place in countries that have emission reduction commitments under the Kyoto Protocol, whereas CDM projects take place in countries that do not.

Source: CDM Watch, *The World Bank and the carbon market: rhetoric and reality*, April 2005, p. 7.

large dams in 2005,[31] the largest of which was the Nam Theun 2 dam in Laos. Comparisons with studies of Brazilian reservoirs suggest that methane from rotting biomass in the dam's reservoir could contribute the same amount to climate change as a natural gas plant generating the same amount of power.[32] The project will also divert 93 per cent of the Nam Theun's flow into the Xe Bang Fai River basin, generating power for Thailand's electricity grid but drastically altering the character of two rivers, displacing thousands of people and disrupting the livelihoods of tens of thousands more.[33] Writing in the *International Herald Tribune* in April 2005, David F. Hales of the Worldwatch Institute described the Nam Theun dam as 'a direct challenge to the recommendations of the World Commission on Dams',[34] whose 2000 report was considered the benchmark for sustainable hydropower development. Patrick McCully of the International Rivers Network has said:

> Instead of pushing the much needed rapid scale-up of investment in new sustainable energy technologies, the World Bank is returning to its bad old ways of pushing big dams. This will neither help the climate nor deliver the affordable decentralised power systems needed to reach the 1.6 billion people without electricity.[35]

The record of the PCF is such that 80 environmental and social justice groups delivered a letter to the World Bank in 2004 calling for its closure. The letter condemned the PCF as 'destructive greenwash'[36] – deliberately promoting environmental programmes to deflect attention from environmentally damaging activities. It said that instead of solving problems, the PCF has 'exacerbated existing human rights violations and furthered environmental destruction.'

## Considering climate change

The World Bank's approach to the climate crisis does indeed seem to involve a major contradiction. While claiming a leading role in the drive to reduce carbon dioxide and the other greenhouse gases that fuel the problem, it is also funding the projects that emit these gases. The frustrating thing about this is that the World Bank could actually play a crucial role in tackling climate change. The Bank could change its investment priorities towards keeping the world's fossil fuel reserves in the ground. It could shift its energy subsidies towards renewable energy that meets the basic needs of the poor while also having less of an impact on the climate.

The Bank has never had any qualms in using policy advice and technical assistance to influence the development paths of debtor countries, and it could incorporate consideration of climate change into this. The World Resources Institute found that that over 80 per cent of the World Bank's publicly disclosed lending in the energy sector from 2000 to 2004 did not consider climate change issues in project appraisals and documentation.[37] Project information documents for a 2002 loan of $454 million for energy sector reform in Brazil made no mention of climate change considerations, despite much of the investment going on infrastructure with relatively long life times, and so with a long-term impact on emissions levels.[38] In 2006 a report to the Development Committee of the World Bank stated that 'there is clearly a need to provide stronger incentives and tools for a better due diligence on climate risks in project design.'[39] The report said: 'Unless new more efficient power plants are installed now, especially in the fast growing G+5, the path of higher carbon emissions from the power sector will be locked in for 40–60 years.'[40]

Another example of where the World Bank could incorporate climate considerations into its work are poverty reduction strategy papers (PRSPs), which countries seeking debt relief under the HIPC (Heavily Indebted Poor Countries) Initiative are required to produce. These countries are some of the poorest in the world, and so are among the most vulnerable to climate change. PRSPs and similar documents could be used, for example, to strengthen the ability of countries to adapt.

Poorer countries would have even more funding available for adaptation if they no longer had to divert their wealth into servicing loans. Already, many of the economic policies the World Bank has imposed on indebted countries have reduced their capacities to pursue policies that reduce vulnerability, even though this is very much needed in the face of climate change. This means that the World Bank could also demonstrate its commitment to action on climate change by responding to a basic demand that has been put to it for some time – cancel unpayable and unfair poor country debts.

Any new adaptation measures need, of course, to be more than a simple add-on to the World Bank's development plans. Antonio Hill of Oxfam says:

> Ultimately, climate adaptation means a fundamental transformation of development practice. The natural tendency to tinker encourages us to seek greater powers of prediction and to incorporate the implications of fore-casted climate impacts earlier in the planning process. These efforts are important but, ultimately, only add-ons to standard development practice, which depends and plans on the basis of certainty and equilibrium. Climate change is fundamentally about degrees of uncertainty and non-equilibrium changes for which no amount of research can help plan. To go beyond development add-ons, effective adaptive measures will need to reinforce approaches tried and tested in environments where people have learned to live with uncertainty. As an example, pastoral

communities in dryland sub-Saharan Africa have long relied on flexible, opportunistic responses to shocks such as moving livestock, and shifting crop mixes and water sources. Development plans need to build in more and better learning to support continuous adjustments at ground-level, and focus more on the adaptive capacity of people themselves rather than pre-determined outcomes. Rather than struggling to deliver more certainty in the face of climate change, adaptation is about enhancing peoples' capabilities to create and pursue different options in the face of uncertain and rapidly-changing climate impacts.[41]

Lisa Schipper, post-doctoral fellow at the International Water Management Institute in Sri Lanka, warns that:

> the popularity of adaptation interventions as a way to address both existing climate-related development challenges and future climate-related development goals (for example, adapting to high rainfall variability now and in the future) risk being overrated. Ultimately, adaptation may become a new mask to hide failed development interventions.[42]

Broader than specific policy documents, funding programmes or individual adaptations, climate change may call for a new model of development. The second report on Africa and global warming from the Working Group on Climate Change and Development said that this should be:

> one in which strategies to increase human resilience in the face of climate change and the stability of ecosystems are central. It calls for a new test for every policy and project, in which the key question will be, 'Are you increasing or decreasing people's vulnerability to the climate?'[43]

The World Bank is in a key position to contribute to this. Will it?

## Box 3.4: Funding

Estimates from the World Bank have suggested that the additional cost of adapting investments to protect them from climate change risks adds up to approximately $4–37 billion each year.

| Item | Amount per year | Estimated portion climate sensitive | Estimated costs of adaptation | Total per year (US$ 000) |
|---|---|---|---|---|
| ODA and concessional finance | | 20% | 5–20% | $1–4bn |
| Foreign direct investment | $160 bn | 10% | 5–20% | $1–3 bn |
| Gross domestic investment | $1,500 bn | 2–10% | 5–20% | $2–30 bn |
| Total international finance | | | | $2–7 bn |
| Total adaptation finance | | | | $4–37 bn |
| Costs of additional impacts | | | | $40 bn (range $10–100 bn) |

Source: Nicholas Stern, *The Economics of Climate Change: The Stern Review*, Cambridge University Press, 2007, p. 442.

The exact figure for protection from climate risk is difficult to calculate. Still, this table gives an idea of the scale of finance needed. It is worth noting that this figure is additional to general costs of adaptation in developing countries, which the World Bank estimates will be tens of billions of dollars. It is also worth noting that, because much of the above is new investment, it doesn't reflect the costs that poor people themselves will bear through loss of livelihoods and assets.

### Sources

Any funding for measures related to climate change will come from a range of sources including private finance, the World Bank and three dedicated funds under the UN Framework Convention on Climate Change (UNFCCC). The creation of these UNFCCC funds was seen as an indication that rich countries were willing to provide poor countries with financial support for adaptation to climate change. For example, the EU and other wealthy countries made a 'political declaration' at COP7 in Marrakech to provide $450 million a year for adaptation.[44]

The funds are outlined below.

### Special Climate Change Fund (SCCF)

This fund covers a number of activities including adaptation, mitigation, technology transfer and economic diversification. It is funded by voluntary contributions, which totalled $45 million by the end of 2006.[45]

## Least Developed Countries (LDC) Fund

This fund is meant to be used by the LDCs to fund only the additional cost of climate change adaptation over and above a development baseline. Its first cost was the preparation of National Adaptation Programmes of Action (NAPAs, see Chapter 1). Pledges and voluntary contributions to this fund amounted to $89 million as of April 2006.[46]

## Adaptation Fund

This is supposed to support 'concrete adaptation projects and programmes in developing country Parties that have become Parties to the Protocol'.[47] Its contributions come from a 2 per cent levy on Clean Development Mechanism deals. Its size therefore depends on the size of the CDM market. The World Bank has estimated that the Adaptation Fund will generate $100–$500 million up to 2012.[48]

## Adaptation Financing Index*

Oxfam estimates that the costs for developing countries of adapting to climate change will be at least $50 billion each year, and far higher if greenhouse gas emissions are not cut soon. It proposes raising adaptation funds through an Adaptation Financing Index. This would be based on the four principles of responsibility, equity, capability and simplicity. Oxfam's approach reveals that 28 countries are both responsible for and capable of financing adaptation in developing countries. The methodology proposed by the Adaptation Financing

Index would see the United States and the EU contributing more than 75 per cent of the finance needed for adaptation, with over 40 per cent from the USA, and over 30 per cent from the EU.

Oxfam says that this funding must not be counted towards meeting the UN-agreed target of 0.7 per cent for aid, as counting it as aid would be an unacceptable inequity in global responses to climate change. It gives the analogy of someone promising to support a child through school, and then breaking the child's bicycle: it would hardly be acceptable to then offer to pay for bike repairs by using money set aside to buy school books.

Much of the funding for adaptation projects at a local level will come from domestic sources. The Stern Review of the economics of climate change said 'Access to insurance and reinsurance services, savings and credit facilities, and flows such as financing for disaster preparedness measures and remittances will also be important to help protect the most vulnerable from climate change'.[49]

* Source: *Adapting to Climate Change: What's Needed in Poor Countries, and Who Should Pay*, Oxfam International, May 2007.

# 4
# Come together as one?

We need to stop this dangerous experiment humankind is conducting on the Earth's atmosphere.

Thomas Loster, climate expert,
Munich Re insurance company[1]

## The story so far

The World Bank is not the only international forum thinking about how to address climate change. Since 1995 the international community has gathered at least once a year at Conferences of the Parties (COPs) to the United Nations Framework Convention on Climate Change (UNFCCC). Like the World Bank, COP meetings tend to be dominated by the richer countries, which are better resourced to participate in, and direct, decision making. Unlike the Bank, this power imbalance is not a structural feature of COPs, as every signatory to the UNFCCC has, in theory, an equal right to participate. This gives COP meetings greater legitimacy as an international forum to tackle climate change.

The UNFCCC's stated objective is to achieve 'stabilisation of greenhouse gas concentrations in the atmosphere at a level that would prevent dangerous anthropogenic interference with the climate system'.[2] Its most high profile agreement is the Kyoto Protocol, which entered into force in 2005, introducing targets for greenhouse gas reduction. The Protocol divides countries into separate groupings based on the principle that the largest share of historical and current greenhouse gas emissions is from industrialised countries. These countries must reduce greenhouse gas emissions by an average of 6 to 8 per cent below 1990 levels between the years 2008 and 2012. Even though it has been a struggle to get to this point, the actual targets are fairly meaningless compared to the level of emissions reductions needed to stabilise the climate (more on this later). The greatest achievement under the UNFCCC so far is probably the fact that it has brought people together, rather than the content of the agreement it has produced. However, given that the climate crisis is pressing, the current negotiations over what happens when the first Kyoto period ends in 2012 are going to have to come up with something far better.

### *Although I hold the mirror, all I see is thee*

The world's biggest carbon dioxide emitter, the United States, signed up to the Kyoto Protocol but never approved it in the Senate. It argued that all countries, not just the industrialised ones, should have binding targets and timetables. This position was most probably motivated by a reluctance to change behaviour, rather than a commitment to fairness. However, a shift in global dynamics since the Kyoto Protocol was first drawn up means that the United States might have a point. For the seats at the table of the world's biggest emitters will soon be occupied by countries that don't currently have a Kyoto commitment to reduce greenhouse gas emissions.

The group of countries known as the developing countries – including India, China, Brazil and South Africa – is expected to become the leading contributor to global emissions by the early 2020s, overtaking the more industrialised countries.[3] The group will be responsible for about 70 per cent of the increase in global carbon dioxide emissions between 2002 and 2030.[4] Heavyweights in the group are China and India, whose emissions increased by 67 per cent and 88 per cent respectively between 1990 and 2004.[5] According to the *Financial Times*: 'Any reductions in emissions achieved by the Kyoto Protocol will look puny when compared with the probable rise in emissions as these economies chase their developed counterparts.'[6]

While this means that traditionally poorer countries do have to be involved in any international framework on climate change, it does not mean that the rest of the world can sit back and wait for China, India and the rest to take action. To begin with, the climate problem was created by the richer countries in the first place: in 1990 (the base year for the Kyoto Protocol), industrialised countries were responsible for 75 per cent of all carbon dioxide emissions and 88 per cent of the emissions that had previously caused climate change.[7] Secondly, one of the driving forces behind industrial development (and thus emissions) in traditionally poorer countries is the rich-country consumer, whose goods are no longer manufactured in his/her own backyard but, under the globalised trading system, come from places where labour and other costs are cheapest. Three little words – 'made in China' – dominate shopping baskets in the world's richer countries. 80 per cent of the companies in Wal-Mart's database of suppliers were Chinese, according to the Worldwatch report *State of the World 2006*.[8] The United States 'saved' 1711 million tonnes of carbon dioxide emissions between 1997–2003 by importing goods from China rather than making them within US borders.[9] Of course the Chinese people benefit from producing goods, so have to take some responsibility for greenhouse gases emitted. But should they take all the responsibility?

Also, many multinational companies may outsource production overseas but still trade their stocks and shares in richer countries. The UK, for example, officially only emits about 2 per cent of global emissions but has companies trading on the London Stock Exchange that are responsible for more than 12 per cent of the world's total emissions.[10] Consumers and globalised corporations with financial and political influence must recognise that their true impact on the global climate is not confined to their coastline.[11]

The choices that China and other rapidly industrialising countries make about their energy sources are, of course, absolutely critical to global efforts to reduce greenhouse gas emissions. However, rich countries, which are also due to remain the highest emitters per person (see Box 4.1), have to change consumption patterns and support moves to cleaner energy in poorer countries, rather than laying blame on those who, quite literally, manufacture their way of life.

**Box 4.1: Carbon dioxide emissions in China, India, Europe, Japan and the United States, 2004, and increase 1990–2004**

| Country or region | Carbon emissions (million tonnes) | Carbon emissions per person (tonnes) | Carbon emissions per unit of GDP (tonnes per million US dollars) | Increase in carbon emissions 1990–2004 (per cent) |
|---|---|---|---|---|
| China | 1,021 | 0.8 | 158 | + 67 |
| India | 301 | 0.3 | 99 | + 88 |
| Europe | 955 | 2.5 | 94 | + 6 |
| Japan | 338 | 2.7 | 95 | + 23 |
| USA | 1,616 | 5.5 | 147 | + 19 |

Source: Worldwatch Institute, *State of the World 2006*, Worldwatch Institute, 2006, p. 9.

## What to do

Taking a multi-stage approach to emissions reductions is one way of getting around the now-outdated division between a group of countries with obligations to reduce emissions and another group with none. One proposal, called the 'South–North Dialogue on Equity in the Greenhouse', suggests dividing countries into six groups representing six different stages.[12] Movement between them would be based on three criteria: (1) historical responsibility, (2) capability (as measured by per capita gross domestic product and the human development index), and (3) potential to mitigate. The proposal links adaptation funding to responsibility for the impacts of climate change based on the 'polluter pays' principle, and includes support for capacity building in developing countries, and education of policy makers and the public.

This South–North Dialogue is just one of many ideas put forward for tackling climate change. While it suggests capping the amount of gases in the atmosphere through fixed emissions targets, other proposals include: targets for emissions in relation to gross domestic product; targets for emissions per person; or different targets for different industrial sectors, as has been used in the EU's Emissions Trading Scheme. Other climate change proposals have rejected the idea of targets altogether. For example, the Coalition for Climate Technology proposed that like-minded countries should cooperate on technology development, rather than negotiating emissions allocations. Some of the technologies under consideration for tackling climate change may seem far-fetched but that doesn't mean they're not receiving government funding. An Associated Press report in March 2007 revealed that NASA has spent $75,000 mapping out the concept of a 'sun shade', which could feature a cloud of small disc-shaped spaceships that go between Earth and the sun and act as an umbrella, reducing heat from the sun.[13] About 16 trillion of these could be launched into space in a series of 20 million rocket launches at a total cost of at least $4 trillion over 30 years.[14] Meanwhile, the National Center for Atmospheric Research (NCAR) – one of the most high-profile climate modelling centres in the United States – spent six weeks in early 2007 running computer simulations of an artificial volcano, which used jet engines, cannons or balloons to get sulphate particles into the upper atmosphere in order to reflect sunlight.[15] NCAR scientist Caspar Ammann pointed out that 'instead of investing so much into this, it would be much easier to cut down on the initial problem',[16] while Stanford University professor Stephen Schneider called such desperate measures 'planetary methadone for our planetary heroin addiction'.[17]

The distraction of ludicrous and expensive ideas at home is one thing. But the United States is also distracting from the Kyoto Protocol by creating an alternative multinational forum. In 2006 the United States joined with Australia, China, India, Japan and the Republic of Korea to launch the Asia-Pacific Partnership. This aims to 'develop, demonstrate and implement cleaner and lower emissions technologies that allow for the continued economic use of fossil fuels while addressing air pollution and greenhouse gas emissions.'[18] Australia's Minister for Industry said: 'While Kyoto puddles on nicely, the real reductions will come from technology.'[19] He may have had a point that the Kyoto Protocol was going nowhere fast, but the Asia-Pacific Partnership is not necessarily going to get to a low carbon world any more quickly. Technological developments may have a role to play in emissions reduction (see Chapter 5), but only when they are part of an overall strategy that includes a limit on emissions. A limit of some sort is needed to provide a context – a structure and a measuring-stick – for decision making.

Decisions over what technologies to adopt are not politically neutral either. As Chapter 5 will outline, there are different social and environmental implications for an energy system based on nuclear technology than there are for one based on renewable technologies.[20] The report from the 2005 conference on avoiding dangerous climate change said that technology:

> has contributed toward a whole host of emerging environmental problems, ranging from indoor air pollution to global climate change. ... At the same time, technology holds the promise to help radically reduce greenhouse gas emissions and other adverse impacts of human activities on a wide range of planetary processes. ... Technology can amplify as well as alleviate adverse impacts of human activities.[21]

## Box 4.2: Stabilisation wedges

Scientists Rob Socolow and Stephen Pacala have suggested that a significant amount of emissions reductions could come from a combination of technologies, where each class of technology would constitute one 'stabilisation wedge'. A wedge is one billion tons of carbon per year of emissions savings by the middle of the century. This 50-year timespan is useful, says Socolow, because it divides the work among generations. It is:

> long enough to allow major changes in infrastructure and consumption patterns, but it is also short enough to be heavily influenced by decisions made today. It is a time frame, looking forward, with

which many businesses are comfortable, and a time frame, looking backward, that is contained in a single human memory.

Socolow's wedges include energy efficiency, carbon capture and storage, nuclear and renewable energy. The energy efficiency wedge, for example, would include avoiding investments in facilities like power plants and apartment buildings that are energy-inefficient or carbon-wasteful.

A wedge could be made up of two million one-megawatt windmills displacing coal power; it could be two billion personal vehicles achieving 60 miles per US gallon (mpg) on the road instead of 30 mpg; it could be capturing and storing the carbon produced in 800 large modern coal plants. Socolow says:

> implementing seven wedges should place humanity, approximately, on a path to stabilising the climate at a concentration less than double the pre-industrial concentration, leaving those at the helm in the following 50 years in a position to drive carbon dioxide emissions to net zero emissions.

He hopes that his proposal would bring about 'a world transformed by deliberate attention to carbon', saying:

> if those alive today bring about the dramatic reductions in carbon dioxide emissions that appear to be our assignment for the next 50 years, the world will be so transformed that the options for the following 50 years will be myriad.

Source: Section VII 'Technological Options', in Hans Joachim Schellnhuber, Wolfgang Cramer, Nebojsa Nakicenovic, Tom Wigley and Gary Yohe (eds), *Avoiding Dangerous Climate Change*, New York, Cambridge University Press, 2006, pp. 347–54.

## Contracting and converging

One of the most high-profile possible frameworks for emission cuts and climate stabilisation is 'Contraction and Convergence'. The Global Commons Institute has proposed this as a global framework based on a commitment to equal rights. 'Contraction and Convergence' requires the international community to initially answer just two questions:

◆   To what level should the concentration of greenhouse gases in the atmosphere contract?
◆   On what date should global per capita emissions converge?

Answers to these questions would give a more robust framework to emissions reductions than the Kyoto Protocol, where figures for emissions cuts are based on what was considered politically possible during the negotiations, rather than what was scientifically necessary.

For the first question, the link between greenhouse gas concentrations and temperature suggests that the answer – the level to which greenhouse gases should contract – depends on what temperature rise is considered acceptable. The UN Convention on Biological Diversity has suggested that an increase in the Earth's average surface temperature beyond 2°C above pre-industrial levels is unacceptable for ecosystems and biodiversity.[22] In itself this temperature increase will bring problems. At 2°C agricultural yields fall, billions experience increased water stress, additional hundreds of millions may go hungry, sea level rise displaces millions from coasts, and regional ecosystems begin to disappear.[23] Beyond the 2°C threshold, the UK's Institute for Public Policy Research (IPPR) says that 'the extent and magnitude of impacts are likely to increase in a way that may widely be considered as being dangerous, and in some cases irreversible.'[24] (For more on possible climate impacts at various increases in temperature see Table I.1 in the Introduction.)

Research by the IPPR suggests that, to have a 'very low to low risk' (a 9 to 32 per cent chance) of going over the 2°C threshold, concentrations of carbon dioxide alone would have to peak at 410–421ppm in the middle of this century.[25]. They would then have to fall to 355–366ppm by 2100.[26] Concentrations of other greenhouse gases such as methane would have to be similarly reduced.[27] Because carbon dioxide stays in the atmosphere for a long time the IPPR says that this translates into global carbon dioxide emissions having to peak in the next decade, before falling to about 70 to 80 per cent below 1990 levels by the middle of the century.[28]

The answer to the first question under 'Contraction and Convergence' would have to be based on the best available scientific information on what risks and impacts match with what temperature change and level of greenhouse gas concentrations. However, the answer is ultimately a political decision as it involves making a judgement over what level of risk is tolerable. Poorer people, whose concerns have often been marginalised in international climate negotiations, would have to be included in this decision as it is they who are most vulnerable to the risks and impacts associated with climate change.

The second aspect of 'Contraction and Convergence' involves setting a date for convergence – a date at which everyone on the

## Box 4.3: Emission pathways

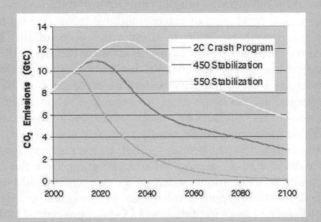

This graph shows three emissions pathways – the path that would see a temperature rise of no more than 2°C, and the pathways for stabilising atmospheric carbon dioxide concentrations at 450 ppm and at 550 ppm.

Source: T. Athanasiou, S. Kartha, P. Baer, *Greenhouse Development Rights: An Approach to the Global Climate Regime that Takes Climate Protection Seriously While Also Preserving the Right to Human Development*, EcoEquity and Christian Aid, 2006, p. 2 (http://www.ecoequity.org/GDRs/GDRs_Nairobi.pdf).

planet has an equal entitlement to emit greenhouse gases. Dates that have been suggested include 2030 or 2050. This 'per capita' approach – where every person, rather than every country, has equal entitlements – recognises that the atmosphere is common property on which everyone depends, and everyone on the planet should have equal rights to use it. An agreement that includes all the world's peoples also recognises that everyone has some responsibility for sticking to a safe limit for emissions, so it would fulfil the United States' requirement that developing countries are given emissions targets.

Under 'Contraction and Convergence', once answers to the two questions above had been found, each country would then be allocated its share of the overall emissions budget. This would come in the form of tradable emissions permits. The cap on overall greenhouse gas emissions would limit the amount of permits available, making these a more robust commodity than under current trading systems. And money raised from trading permits could finance clean energy schemes or adaptation.

While 'Contraction and Convergence' has political integrity, as it offers some justice through equity and gives real entitlements to poorer countries, is it politically realistic? It allows poorer countries to increase their emissions but only if richer countries cut theirs. Is this really a possibility? Andrew Simms, policy director of the new economics foundation, believes it may be. He has compared the rationing of emissions that would take place under 'Contraction and Convergence' to the rationing of general goods in the United Kingdom during the Second World War. The UK saw a 95 per cent drop in the use of motor vehicles and 31,000 tonnes of kitchen waste saved each week – enough to feed 210,000 pigs.[29] At that time, rationing was generally accepted, as it was believed to be for the good of the country. 'All these problems have been wrestled with before and, to a degree, have been overcome,' says Simms. 'But they have been overcome only when the urgency and necessity is generally understood.'[30] This time the enemy is not Hitler or another country but a hostile atmosphere 'that needs to be disarmed because of its increasingly violent arsenal of droughts, floods and storms.'[31] Simms points out that the world's poorest countries have been reshaping their economies for years under the World Bank's Structural Adjustment Programmes in order to pay service on foreign debts. He says: 'it would be a shameless double-standard now to suggest that we in the rich world, using the targets that "Contraction and Convergence" will give us, can't work within the framework of "sustainability adjustment programmes" to balance our ecological budgets.'[32]

One tool that may help us balance our ecological budgets is individual carbon trading – also known as personal carbon allowances, domestic tradable quotas, personal carbon rations, and carbon credits.[33] If every person was given a set emissions allowance within an overall framework of a limited (and reducing) number of allowances, carbon trading might actually help reduce emissions. And the allocation of an emissions quota does not have to mean lower living standards, as some might have described the Second World War rationing regime. It means a more efficient use of resources in order to achieve the standard of a cleaner and stable environment, with a chance of supporting future generations.

Under an individual carbon trading scheme, those who needed or wanted to emit more than their individual allowance could buy allowances from those who manage to live within their carbon means. Individuals could have something like a carbon credit card to 'swipe' and spend their allowances as they buy things. Those who felt the system was too complicated could simply cash in their allowances and pay higher energy prices year-round as a tax.

A UK government department initial analysis of the concept of individual carbon trading said that while more research is needed to find out whether such a scheme really would trigger change in behaviour it would, at least, 'enforce and incentivise individual responsibility amongst a population which has so far appeared unable and/or unwilling to constrain its collective urge to drive, fly and consume more electricity.'[34]

## Box 4.4: A perspective on equity

The North typically views equity issues in terms of fair allocation of emission reduction targets, while the South sees the key equity questions as pertaining to responsibility for climate change and experience of the negative impacts from climate change. These differing views on climate change equity are related to different perceptions about how climate change may affect human security. In the North, climate change is not seen as critical threat to human security, but instead is characterised as an environmental pollution problem that can be addressed through lifestyle changes and pollution control policies. In the South, by contrast, climate change is considered a life-threatening human welfare problem which circumscribes the potential for development.

Source: Karen O'Brien and Robin Leichenko, *Climate Change, Equity and Human Security* – a paper at the workshop *Human Security and Climate Change*, Asker, near Oslo, 21–23 June 2005, p. 10.

### All aboard?

It is not just the attitude of the industrialised countries and their citizens that may pose a problem for the adoption of 'Contraction and Convergence'. The developing countries that are on course to be the world's biggest emitters may not be so happy about accepting a limited emissions budget, as their past record features low emissions, and their future plans feature industrial development. Proponents of 'Contraction and Convergence' argue that their scheme gives these countries enough permits to allow them space to develop. However, some argue that, while richer countries still have a share of the emissions cake, the slice left for developing countries can never be big enough. In a posting to an Internet discussion list Tom Athanasiou of EcoEquity suggested that 'the Northern bankrupting of the greenhouse-gas budget' will put any poor country in a position where it 'is forced to radically curtail its emissions ... long before it has reached a (nationally averaged) level of wealth even vaguely comparable to that which the Northern countries enjoyed when they first started to curb their emissions.'[35]

In a paper written with Paul Baer of EcoEquity and Sivan Kartha of the Stockholm Environment Institute, Athanasiou identified a core tension between poorer countries' development aspirations and the need to do something about climate change. This was that: 'as economies are now structured, and as development is still conventionally envisioned, ending poverty unavoidably means vastly improved access to energy services and rising carbon emissions.'[36] Athanasiou, Baer and Kartha compared two scenarios: (1) the emissions path that poor countries would have to follow as part of a global plan to limit the temperature rise to no more than 2°C, with (2) the emissions path that the IPCC has predicted poorer countries will follow under a scenario of optimistic assumptions about equity and economic growth. They found that the precautionary path of (1) was radically inconsistent with the optimistic predictions of (2) (see Box 4.5). To meet the 2°C limit, emissions in poorer countries would have to leave their projected path quite soon, and drop dramatically by 2020.

The three authors posed the question: 'What manner of climate regime can enable such a rapid emissions decline at the same time as the South continues, and even steps up, its fight against poverty?' Their answer is an approach that builds on

## Box 4.5: Emissions budget for developing countries

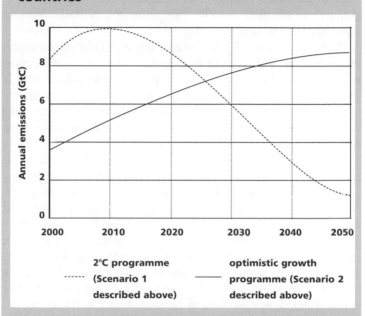

2°C programme (Scenario 1 described above)

optimistic growth programme (Scenario 2 described above)

Scenario (1) plotted against scenario (2). Emissions in richer countries are assumed to drop to zero in 2020 so that the South's budget reflects the entire global emissions budget. Without this assumption, Southern emissions would have to start their decline well before 2020.

Source: T. Athanasiou, S. Kartha and P. Baer, *Greenhouse Development Rights: An Approach to the Global Climate Regime that Takes Climate Protection Seriously While Also Preserving the Right to Human Development*, EcoEquity and Christian Aid, 2006, p. 3 (http://www.ecoequity.org/GDRs/GDRs_Nairobi.pdf).

the idea that global emissions should contract while per capita emissions move towards convergence, while also aiming to 'break the global impasse over developmental equity in a climate constrained world'. This approach is based around 'Greenhouse Development Rights'.[37]

The Greenhouse Development Rights (GDRs) approach establishes an indicator of responsibility that reflects a country's contribution to the climate problem – its carbon debt. It also includes an indicator of a country's capacity – its ability to dedicate resources to the problem. These indicators account for inequalities within countries by calculating them in a disaggregated way, rather than taking national per capita averages. This is sensitive to income differences within countries, and accounts for the fact that a country could have a wealthy minority while also having an impoverished majority.

The GDR approach starts with all countries being obliged to reduce their emissions to an agreed level, beyond which industrialised countries would have different obligations from developing countries. It discards the conventional Kyoto categories, establishing instead a development threshold reflecting a level of socio-economic development to which all countries are entitled. This threshold is measured by a capacity indicator that includes, but is not necessarily limited to, per capita income. Those countries whose capacity indicator exceeds the development threshold would have to pay for the low-carbon development needed to meet the global mitigation shortfall. Countries with greater responsibility and capacity would have to pay for a correspondingly larger proportion of the global mitigation shortfall.

If GDRs were adopted internationally, those countries whose capacity indicator falls below the development threshold would not have to contribute to meeting the global mitigation shortfall. Instead, they would have to put resources directly towards human development. Once a country reaches a development threshold it would, by definition, have enough capacity to start paying for the global mitigation shortfall, though its initial mitigation obligations would be small. Until that time, development would be its proper priority. Finally, all countries would be obliged, in proportion to their obligation indicator, to pay to meet the adaptation funding shortfall.

The GDR approach aims to transfer resources from the wealthy countries in order 'to provide the resources necessary to allow society, collectively, to transition to clean, efficient, low-carbon economies, and to meet a precautionary global emission trajectory'.[38] It recognises that poorer countries will refuse to pay the additional costs of low-carbon energy technology until their most pressing human development needs have been met: 'the problem of human development, in other words, is intrinsic to the problem of negotiating a global climate regime.'[39]

Whatever climate regime is adopted when the first Kyoto phase ends in 2012, poorer countries need more practical support in order to participate in a meaningful way. In *'On Behalf of my Delegation ...' A Survival Guide for Developing Country Climate Negotiators*, Joyeeta Gupta pointed out that while some countries are able to send delegations consisting of dozens of lawyers and diplomats to the Kyoto negotiations, other countries could only afford to send single-person delegations with limited experience and expertise on international negotiations on climate change.[40] Jouni Paavola of the UK's Centre for Social and Economic Research on the Global Environment said that: 'the Convention's general working procedures, small delegations and lack of resources to support negotiation teams continue to hold back the effective participation of developing countries at the international level.'[41] Poorer countries therefore need improved capacity to take part in institutional negotiation, including measures to improve the negotiating skills of national representatives who take part in international climate discussions. Limitations on participation extend beyond the negotiating rooms. If they are going to take part in international emissions regimes, poor countries need to improve their capacity to measure emissions and assess the impacts of climate change.

## Box 4.6: Change happens

The labels of 'victim', 'hero' and 'villain' can be very distracting in the climate change debate. Poorer countries that were traditionally 'victims' – subject to the impacts of climate change yet not having any voice in decision making – look on course to become 'villains' as their emissions begin to climb. One of these, China, still has massive internal divisions between those who are rich and those who are poor. The latter are unlikely to contribute much to the country's 'evil' impact on the climate or to benefit from the country's increasing industrialisation. As for heroes, the former UK Prime Minister Tony Blair toured the world 'heroically' making pronouncements about the importance of tackling climate change yet failed to pursue policies that would have set the UK on a low-emission path.

The greatest climate change 'villain' to date has surely been the United States – the world's biggest emitter and the country most reluctant to admit there is a problem, let alone do anything about it. But amongst US citizens there are signs of change – a climate change movement whose potential is ignored every time the country as a whole is packaged up by its critics as 'villain'.

A report on the 2005 UN climate conference in Montreal:

the multitude of people and organisations from the United States roaming the premises of the conference provided the distinct feeling that the Bush administration was even isolated within its country. ... Dozens of mayors from all over the United States, hundreds of environmental organisations, green business associations, religious groups and also the Inuit from the Arctic all demanded that the United States support effective climate policy and not obstruct progress of the Kyoto Protocol. Many went as far as calling on the rest of the world to go ahead without paying too much attention to the current US administration.'[1]

Many US states are introducing controls on greenhouse gases and are introducing policies that support the development and deployment of clean energy technologies. Even US businesses have recognised the market potential in renewable energy and energy efficiency: the chief executive of Cinergy Corp. told a congressional hearing that 'While the world is deploying leapfrogging technology in an effort to deal with climate change, the US lags sorely behind.'[2]

Sources
1. Bettina Wittneben, Wolfgang Sterk, Hermann E. Ott and Bernd Brouns, *In From the Cold: Climate Conference in Montreal Breathes New Life into the Kyoto Protocol*, Wuppertal Institute for Climate, Environment and Energy, 2005, p. 22.
2. *BusinessWeek online*, 27 June 2005.

## The good, the bad and the urgent

The good news is that, as the above examples of 'Contraction and Convergence' and Greenhouse Development Rights show, in theory it is not impossible to draw up an international agreement that leads to a stable climate, while at the same time generating funding for adaptation measures. Similarly, Andrew Simms' analogy of a country's behaviour when faced with a wartime situation illustrates that it is possible to make the scale of changes needed in a very short time, if motivated enough to do so.

The bad news is that, in practice, it is going to be incredibly hard to get any effective international agreement on climate change. Although an agreement based on binding emissions targets would be the most practicable route to a stable climate, the principle of binding targets remains under debate. Also the character of international climate negotiations is still one of self-interested horse-trading, with policy makers talking about the potential of severe impacts from climate change but still failing to take the necessary actions. They may have signed up to Kyoto targets but that doesn't mean they've got much prospect of reaching them. For example, in 2006 Canada's Environment Minister said that the country would have to ground every train, plane and car to have any chance of meeting its Kyoto target, as its emissions were 35 per cent above the level it promised to reach under the Protocol.[42]

After the 'good' and the 'bad', the 'urgent' is the pace at which change has to take place if a climate catastrophe is to be averted. A crisis of sorts is already inevitable. James Hansen, director of NASA's Goddard Institute for Space Studies, has said: 'I think we have a very brief window of opportunity to deal with climate change ... no longer than a decade at the most.'[43] Figures quoted on page 97 from the

Institute for Public Policy Research echo this statement, suggesting that greenhouse gas emissions have to peak in the next decade. That doesn't mean there is a decade left in which to make decisions – it means there is a decade left in which rising emission figures have to reverse direction. For good. That's quite a daunting task.

# 5
# It's an issue of energy

The choices we make today cast huge shadows into the future, defining the ambition we can reasonably hold for prosperity and stability in the years to come.

Aaron Cosbey, Warren Bell and John Drexhage, International Institute for Sustainable Development[1]

## Increasing energy use

At the heart of any attempt to tackle climate change lies a question over what to do about energy. Energy activities are the largest source of greenhouse gas emissions. And, although greenhouse gas emissions need to go down, energy use is on the up. The International Energy Agency (IEA) predicts that energy use will increase by 60 per cent by 2030,[2] and 85 per cent of this is expected to come from fossil fuels.[3] Oil's share of energy consumption is predicted to drop by just 1 per cent between 2002 and 2025.[4] Consumption of gas is expected to increase by about 70 per cent during this time.[5] Although use of non-fossil fuels is expected to grow, their share of total electricity generation is expected to stay at a similar level to today's. Much of the growth in renewable energy is expected to come from large dams that, as Chapter 3 mentioned, bring problems of their own. Overall, these predictions seem to come from a different world. If they were to come true in our world they would guarantee climate catastrophe.

## Box 5.1: Fossil fuel use

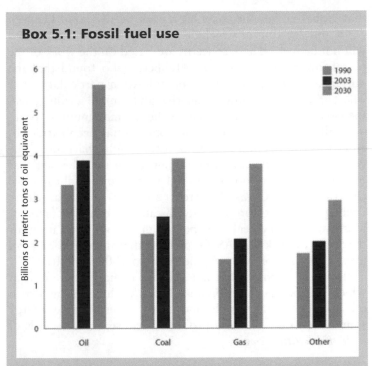

Fossil fuels have been the fastest growing source of energy.

Source: The World Bank Group, *2006 World Development Indicators*, 'Environment', Figure 3e: Use of fossil fuels continues to rise faster than that of other sources of energy.

## A pincer movement at the peak

The above, however, are predictions rather than definite paths of energy use for the coming decades. The IEA itself has said that, although demand for oil is due to rise, 'it's unclear where the supply will come from'.[6] For concerns have been raised that global oil supplies may hit a peak at any point between 2005 and 2037, after which demand will outstrip supply and will not be met.[7] These concerns are based on the work of the petroleum geologist M. King Hubbert, who found that the output of an oil well or reservoir follows a particular curve. The curve sees oil production rise after initial drilling, and then lose momentum as output reaches its maximum or 'peak' – usually when about half of the total oil has been extracted. After this, production falls at an increasingly sharp rate. The website *transitionculture.org* has suggested that the fact that oil companies are looking for oil in ever-more inhospitable environments such as the Arctic and the deep sea, where drilling faces extraordinary technical and other difficulties, illustrates that the peak may be in sight.[8] The Association for the Study of Peak Oil (ASPO) predicts that the world oil peak may occur as soon as sometime between 2008 and 2010.[9]

'Peak oil' is important not because it marks the point at which oil runs out, but because it marks the point at which oil as a cheap, accessible and easily transportable energy source runs out. This is a problem because, in the words of *transitionculture.org*:

> We have created a society that depends very largely on a particular kind of energy. We are dependent on oil because of its versatility, liquidity – which makes it easy to move around – and also because we can make so many things out of it, including plastics and asphalt for our roads. We won't be powering the vast fleets of international air transport on wind power. And we won't be repairing roads with sunbeams.[10]

After the peak is reached there will still be oil in the ground, and in less conventional forms such as tar sands and oil shale. There will be enough oil available to continue damaging the climate. This oil will just be increasingly difficult, expensive and energy-consuming to extract. According to a 2005 report commissioned by the US Department of Energy:

> The world has never faced a problem like [the peaking of world oil production]. Without massive mitigation more than a decade before the fact, the problem will be pervasive and will not be temporary. Previous energy transitions were gradual and evolutionary. Oil peaking will be abrupt and revolutionary.[11]

## Low-carbon routes

So, even if climate change did not mean unprecedented economic, social and political costs, problems with the supply of oil mighty bring them anyway. The two situations will see different responses; for example, supporters of continued fossil fuel use may take concern over peak oil as an argument for new frontiers in oil exploration – exploitation of tar sands or drilling in Alaska. Both climate and peak oil concerns, however, highlight the need for greater energy conservation and energy efficiency and a switch to low-carbon technologies. Both make it necessary for this to happen very soon and on a very large scale.

An obvious technology switch is a move to greater use of renewables, in the form of wind, solar, tidal, small hydro and geothermal power. Already, hot water and space heating for tens of millions of buildings is supplied by solar, biomass and geothermal. Solar thermal collectors alone are used by approximately 40 million households worldwide.[12] By 2005 at least 43 countries, including ten developing countries had put in place an important catalyst – a national target for renewable energy supply.[13] The potential for renewables to underpin changes in the production and consumption of energy was spelt out in a

2007 study by the European Renewable Energy Council (EREC) and Greenpeace. This report, *Energy [R]evolution: A Sustainable World Energy Outlook*, says that renewable energy, combined with efficiencies from the 'smart use' of energy, can meet half of the world's energy needs by 2050.[14] This is more optimistic than the International Energy Agency, which predicts that the share of renewables would gain only fractionally by 2030, to 13.7 per cent of world energy demand.[15] The report noted that the time was right for a shift in energy production as many power plants in richer countries will come to the end of their technical lifetime in the next decade and will need to be replaced, while many power plants in poorer countries are still

## Box 5.2: Projection of global renewable electricity generation capacity under the Energy [R]evolution scenario (in MW)

|  | 2003 | 2010 | 2020 | 2030 | 2050 |
|---|---|---|---|---|---|
| Hydro | 728,000 | 854,800 | 994,190 | 1,091,490 | 1,257,300 |
| Biomass | 48,030 | 110,000 | 211,310 | 305,780 | 504,610 |
| Wind | 30,280 | 156,150 | 949,800 | 1,834,290 | 2,731,330 |
| Geothermal | 10,170 | 20,820 | 40,780 | 70,380 | 140,010 |
| PV | 560 | 22,690 | 198,900 | 727,820 | 2,033,370 |
| Concentrating solar power | 250 | 2,410 | 29,190 | 137,760 | 404,820 |
| Ocean energy | 240 | 2,250 | 13,530 | 28,090 | 63,420 |
| Total | 817,000 | 1,169,120 | 2,437,700 | 4,195,610 | 7,134,860 |

Source: Sven Teske, Arthouros Zervos and Oliver Schäfer, *Energy [R]evolution: A Sustainable World Energy Outlook*, Greenpeace and the European Renewable Energy Council, January 2007, p. 42.

on the drawing board. The *Energy [R]evolution* report said: 'renewable energy is not a dream for the future – it is real, mature and can be deployed on a large scale. ... All that is missing is the right policy support.'

However, while a wholesale shift to renewable energy may be sensible, it is not entirely likely. Two of the world's most energy-hungry countries – China and India – sit on vast reserves of the fossil fuel coal. Over half of India's total primary energy consumption and two-thirds of China's comes from coal.[16] A 2002 study by China's Energy Research Institute suggested that, in the best-case scenario, the share of coal in China's primary energy mix would decrease from 70 per cent to 60 per cent between 2000 and 2020.[17] In India, half the population lacks reliable access to basic electricity services and many industries have an unreliable power supply, so power generation is likely to remain a top priority. The country's Committee on Integrated Energy Policy has said that coal must keep its current share in India's overall energy mix for the next 25 years if the country is to sustain economic growth at over 9 per cent per year.[18] Already, India's coal-fired power generation capacity is expected to more than double by early next decade.[19]

On one hand, neither China nor India will limit its economic development by leaving coal in the ground in order to make up for a problem that the rich world created. On the other hand, there is simply no way that climate chaos can be avoided without China and India finding a different way to fuel their development. '[Coal] is the key to the door of the overall technology shift needed [to curb] climate change,' said Tom Burke, visiting professor at Imperial College in London.[20]

## Clean coal?

There are technologies that may allow coal-rich countries to make the most of their domestic energy source. None are 'the answer' to climate change. None are emission free. But emissions reductions have to start taking place before fossil fuels have been phased out so, in the meantime, these technologies may make some reduction in greenhouse gas emissions from coal: they may help to bridge the gap until renewables are more widely available.

The first of these coal technologies is the integrated gasification combined-cycle (IGCC). This produces so-called 'clean coal', where crushed coal is heated and pressurised to turn it into a gas. Carbon dioxide and smog-causing pollutants are then filtered from the gas before it is burned.

The process of 'underground coal gasification' may also avoid some of the environmental destruction involved in coal extraction. The writer George Monbiot describes this process as follows:

> Holes are drilled into a coal seam and air and steam are pumped into it. The coal is 'gasified', releasing methane and hydrogen, both of which can be burnt in power stations. Carbon can be extracted from the gases either before or after they are burnt. The technique requires no major excavations, no tailings or slag heaps, and no children breathing dust in narrow galleries.[21]

Another bridging technology is carbon dioxide capture and storage (CCS). This involves capturing carbon dioxide and transporting it by tanker or pipeline to be stored underground in depleted oil reservoirs, depleted natural gas fields or deep saline aquifers. This is already happening at the Sleipner gas field in the Norwegian North Sea, where carbon dioxide is removed from the gas and pumped into a highly permeable sandstone. The IPCC has said there is considerable potential for CCS, particularly if applied in the electricity sector, suggesting that such systems could provide between 15 and 55 per cent of the cumulative mitigation effort until 2100.[22]

But many questions still remain about the environmental risk, safety and costs of CCS systems. The International Energy Agency has said the costs are so great that the technology 'will not be applied on a large scale without strong government support'.[23] Whether or not CCS can be deployed on a scale that would make it economically viable, in the time that's left to turn around emissions figures, remains to be seen. Charlie Kronick, senior policy adviser at Greenpeace, said that the technology was at least ten years off, while action could be taken now to reduce emissions through, for example, energy efficiency and renewables.[24]

CCS raises another unexpected question. This is, will it really reduce emissions? For not all the technology's supporters are committed to less use of fossil fuels: some want to use CCS to access more. If carbon dioxide is pumped into an oil field the pressure in the field increases, making oil that was previously hard to extract more accessible. In a test project in Canada, the process increased an oil field's production by 10,000 barrels a day.[25] The US Department of Energy believes that it could boost its oil reserves fourfold through advanced injection of carbon dioxide into depleted oilfields.[26] And in Scotland the energy company BP has considered extending the life of the Miller oil field under the North Sea by at least 15 years by using CCS at a power plant in Peterhead, Aberdeenshire.[27]

Without a political framework that aims for a low carbon world, CCS may be used to make more oil available, contributing to higher, rather than lower, greenhouse gas emissions.

## False promises

Although there are issues around CCS that still need to be resolved it could, given the right investment and political

## Box 5.3: Wider problems with coal

Beyond its impact on the global climate, coal is a polluting fuel at a more local level. Consumption of coal is the main cause of China's urban air pollution and acid rain. Of the world's 20 cities with the most polluted air, 16 are in China, and more than 80 per cent of Chinese cities have sulphur dioxide or nitrogen dioxide emissions above the World Health Organisation's threshold.[1]

The construction of China's new power stations has led to social as well as environmental problems. For example, in 2005 up to 20 people died in a village in Guangdong after Chinese militia opened fire on demonstrators opposing a coal-fired power plant. Local residents said they had not been compensated for loss of income and land caused by the construction of the plant, or for the likely deterioration in the air and water quality.[2]

Sources
1. Worldwatch Institute, *State of the World 2006*, Worldwatch Institute, 2006, p. 7.
2. *Guardian*, 12 December 2005.

framework, possibly contribute towards a more manageable level of greenhouse gas emissions. This is not the case for at least two of the other technologies that have been put forward as a solution – nuclear power and biofuels.

According to US President George Bush, nuclear power is 'safe and clean' and it generates 'large amounts of low-cost electricity without emitting air pollution or greenhouse gases'.[28] This is not the case. All energy systems involve fossil fuel use at some point in their overall life cycle. Wind turbines, solar panels and tidal barrages, for example, have to be manu-factured somehow. The development of energy systems based on nuclear power is particularly fossil fuel-intensive. Nuclear power needs fossil fuels for uranium mining, enrichment and transport. It needs them for the construction and decommis-sioning of facilities. It needs them for the processing, transport and storage of radioactive wastes. 'Embracing nuclear power as a remedy to global warming is like taking up heroin to avoid an addiction to crack,' said Alice Slater, President of the Global Resource Action Center for the Environment.[29]

Worldwatch's *State of the World 2006* report found that, even if the approximately 30 new nuclear plants that both the Indian and Chinese governments plan to build over the next two decades are built, neither country will be getting even 5 per cent of its electricity or 2 per cent of its total energy from nuclear power in 2020.[30] The lead time involved in building new facilities and producing nuclear power means that it will contribute little to reducing emissions in the crucial window of the next decade. Meanwhile, problems that have always been associated with nuclear power – waste, cost and vulnerability to terrorism – have not gone away.

Biofuels – energy crops that have been grown either for biomass burning (to produce heat and energy) or for transport fuel – are a similarly short-sighted 'solution' to climate change. Their supporters argue that biofuels produce less greenhouse gases than oil, and any emissions they produce when burned are not necessarily additional because the plants have absorbed carbon dioxide while growing. As with nuclear power, however, a look at the full life cycle of biofuels reveals considerable greenhouse gas emissions. These come from the processes of refining and distilling, as well as from transport, the use of farm machinery, and fertiliser production.

However, those emissions pale into insignificance beside the emissions generated right at the beginning of the biofuel life cycle. The key growing regions for biofuels such as oil palms and soya are Indonesia, Malaysia and the Amazon. Here, vast forests that act as vital carbon sinks and provide habitat for a range of species are being felled in order to clear land for the biofuel crops. Dr Ian Singleton, Scientific Director of the Sumatran Orangutan Conservation Programme said:

> Huge areas of forests, habitat for the remaining orang-utan on Borneo and Sumatra, are being destroyed in the race between Malaysia and Indonesia to become the world's biggest supplier of palm oil. Conversion to oil-palm estates completely eradicates forests, and annihilates orang-utan populations within them, and those of countless other species.[31]

The development of oil-palm plantations was responsible for 87 per cent of deforestation in Malaysia between 1985 and 2000.[32] In Asia, some of the forests being cleared for biofuels are swamp forests, which grow on peat. When the peat is dried out in order to create a palm plantation, it oxidises and releases massive amounts of carbon dioxide. In Brazil, where President Lula said that biofuel 'is significantly less polluting than conventional petroleum-based diesel',[33] Giulio Volpi, co-ordinator of the WWF's climate change programme for Latin America and the Caribbean, said:

> Brazil is set to produce most of its biodiesel from soya beans, which have virtually no advantage over conventional fuels in terms of overall greenhouse gas emissions, let alone the millions of hectares of tropical forest that have been cleared for large-scale soya plantations.[34]

'Travelling in a car fuelled by biodiesel seems like a great, environmentally-friendly thing to do,' said Ariel Brunner of the non-governmental organisation BirdLife. 'However, if the biodiesel has come from soya planted in the Brazilian Amazon or palm oil from Indonesia, the green consumer is likely to be unwittingly driving another nail into the coffin of the world's great ecosystems.'[35]

Social problems also follow the establishment of biofuel plantations. Jose Taveira da Silva, a rural union leader in Brazil, said:

> the soya growers are following the loggers into land that was forested, and there have been cases of small farmers suffering threats or actual violence when they refuse to sell. Small farmers produce for their own family and the local markets. Soya is geared to exports. Does this improve the lives of the local people?[36]

In Indonesia, Friends of the Earth described the oil plantation business as the country's most conflict-ridden sector.[37] Plantations are often established by force on land traditionally owned by indigenous peoples, and plantation development has repeatedly been associated with violent conflict. Between 1998 and 2002 alone, 479 people in Indonesia were reported as having been tortured in conflicts defending community rights, and dozens of people have been killed in land-tenure disputes.[38]

Biofuels have to be grown somewhere, and even if they are grown on land where no peat has been drained or valuable forest destroyed they are likely to be in direct competition for land and water with food crops. Alexander Mueller, assistant Director General of the UN Food and Agriculture Organisation, called biofuels one of the three major challenges for farming, suggesting that rising production of biofuels from crops might complicate UN goals of ending hunger.[39] A 2007 report from the Reuters news agency told how soaring US demand for ethanol – produced from crops like maize and sugar cane – sent maize prices to their highest level in a decade, impacting on Mexico, where tens of thousands took to the streets after the price of (maize-based) tortillas tripled.[40]

## Energy security

When Thomas Edison had the world's first power station built in New York in 1882, he could surely not have envisioned the complex power factories we have come to rely on today. This was not just because his imagination for technical possibilities was limited by the experiences of his time. Such large, centralised power systems may not have crossed his mind because they pose problems that are still relevant – the extensive infrastructure they require makes them inefficient and very expensive to set up. These problems mean that, as Cowan Coventry of Practical Action said: 'Grid extensions will reach less than half the two billion waiting for access [to electricity] within the next thirty years.'[41] Yet energy is essential for development. A new economics foundation report, *The Price of Power: Poverty, Climate Change, the Coming Energy Crisis and the Renewable Revolution*, says:

> Ready access to clean and reliable energy can soften several hardships for poor communities, such as cooking, and the need for heating, cooling, water pumping and sanitation and by replacing other expensive, polluting fuels. It can mean electricity in schools and that children can do their homework in the evening. It can provide higher efficiency for small-scale industry, as well as release women from tedious and time-consuming tasks, such as fuel wood collection, countering the costs of gender inequalities.[42]

Access to energy also means less reliance on traditional

biomass for cooking and heating – a reliance that contributes to up to two million deaths each year through indoor air pollution.[43] The *Price of Power* report says that: 'Getting the answer to the energy question right will determine the success or failure of international efforts to meet the millennium development goals.'[44]

Edison's original vision was a decentralised energy industry with power generated and delivered close to where it was to be used. This vision still holds great potential, for rural communities in particular. Typical decentralised energy technologies include renewables such as solar photovoltaic systems, small hydro, wind power and geothermal production. These are already proving effective not just for meeting basic needs but for developing adaptations to climate change. For example, in the drought-prone region of the Sahel in Africa, where both access to water and quality of water are major problems, solar photovoltaics have been used to power water pumps that have contributed to irrigation, livestock needs and village water supplies.[45] *The Price of Power* estimates that approaching Africa's energy needs through a mixture of renewable energy technologies would cost considerably less than $50 billion.[46]

*The Price of Power* says that, as well as being reliable, decentralised renewable systems give people more control over the electricity that powers their lives:

> Not only can renewable energy provide a clean, flexible power source for homes, schools and hospitals, at the micro-to-medium scale it has huge potential to create meaningful and useful jobs. By literally taking control of their own power supply, marginalised communities and marginalised people within communities can also be empowered. In this way renewable energy provides immediate improvements to people's lives, but it also

## Box 5.4: Energy spending in sub-Saharan Africa

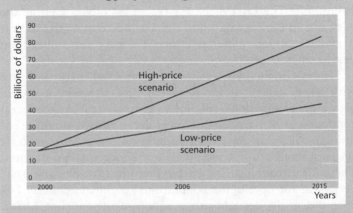

Christian Aid estimates that if sub-Saharan Africa grows by 7.1 per cent a year – the amount the UN says will be required for it to achieve its Millennium Development Goals by 2015 – and continues to use oil as one of its primary fuels, it will be pumping an extra 76 million tonnes of carbon a year into the atmosphere by 2015.

On the other hand, says Christian Aid, providing solar electricity to a village of 50 households would cost an average of $25,000. Assuming that the average household size in sub-Saharan Africa is five people, this works out at a cost of about $100 per person. When multiplied by the number of people in sub-Saharan Africa without electricity – about 500 million – providing solar electricity works out at about $50 billion. This figure compares favourably with the amount the region is likely to have to spend on oil over the next decade.

Source: Christian Aid, *The Climate of Poverty: Facts, Fears and Hope*, 2006, p. 17 (http://www.christian-aid.org.uk/indepth/605caweek/caw06final.pdf).

gives a key to a different roots-up model of human development in which people are more in control of their own lives and livelihoods.[47]

## Box 5.5: This works

Life in the Indian village of Khadakwadi was transformed with the installation of low-cost, energy-efficient lamps powered by the sun. 'Children can now study at night, elders can manage their chores better,' said a village resident. 'Life doesn't halt anymore when darkness falls.'

With no extra cost after a $55 installation, solar energy lights a lamp that uses LEDs – light emitting diodes – that are four times more efficient than an incandescent bulb.

'This technology can light an entire rural village with less energy than that used by a single conventional 100 watt light bulb,' says Dave Irvine-Halliday, a professor of electrical engineering at the University of Calgary, Canada.[1]

Some communities in the Amazon region in northeast Peru find it impossible to provide certain basic services, such as vaccine and medicine preservation, proper education and communication facilities. Stomach and skin diseases are widespread, as well as contagious illnesses such as hepatitis and tuberculosis.

Despite the general hardship, the Amazon communities have a major energy source right next to them – the river. Using the power of the river El Paraiso, the local community developed a river-current turbine. Due to its success, other neighbouring communities requested support to do the same. Two communities now have improved basic services, with a vaccine refrigerator, lighting, health centres, education centres and other community priorities.[2]

A team of experts in Lao PDR has been training small companies to become village energy service specialists, to work in areas not due for electricity grid connection.

These electricity service companies offer a choice of electricity supply technology, including solar home systems. In the case of villagers choosing solar home systems, the equipment is bought with monthly payments over a period of several years. The solar panel then becomes an important economic asset to poor families, since it retains high re-sale value. The villagers choose a village electricity manager to provide technical support over this period and beyond.

In one village, access to solar power has increased incomes by allowing net-mending to take place at night, and by allowing the charging of batteries used for fishing and for hunting frogs at night.[3]

Every day, many thousands of corn tortillas are sold in three major cities in Nicaragua.

They are produced and sold by over 190,000 house-holds and represent a common way of earning an income for single mothers and families where the husbands are unemployed.

The tortillas are generally cooked on traditional woodstoves, which means working for hours over an open wood fire, exposed to heat, smoke and flames. It leads to severe eye irritation, respiratory complaints and other diseases related to indoor air pollution. But there are solutions that could have huge health benefits for the numerous tortilla producers, such as a modern woodstove known as the Ecostove.

Apart from eliminating indoor air pollution by nearly 100 per cent, increased woodstove efficiency also saves 50 per cent in fuelwood costs and minimises deforestation. A pilot survey has shown that the Ecostove can also reduce the emissions of greenhouse gases by 35 per cent.[4]

Sources
1. *Christian Science Monitor*, 3 January 2006.
2. Andrew Simms, Julian Oram and Petra Kjell, *The Price of Power: Poverty, Climate Change, the Coming Energy Crisis and the Renewable Revolution*, New Economics Foundation, 2004, p. 21.
3. *Tiempo* Issue 55 April 2005, pp. 22–6.
4. Simms et al, *The Price of Power*, p. 26.

## Generating force

The fact that climate change is already taking place is adding momentum to the development of different energy systems. In 2005 China's Vice-Minister of the Environment Pan Yue said that: 'This [Chinese economic] miracle will end soon because the environment can no longer keep pace.'[48] One estimate predicts that China will lose two-thirds of its glaciers by 2050, intensifying water shortages and putting at least 300 million people at risk.[49] Greenpeace has said:

Climate change is causing a cocktail of environmental effects at the Yellow River source that threaten an ecological breakdown. When you see the empty wells,

bridges over nothing but dry dirt, cracked ground
where there should be lakes, bare rock and sand where
it was once healthy grassland you know something is
seriously wrong.[50]

Climate change is also predicted to reduce Chinese food production by 10 per cent between 2030 and 2050.[51] If climate change
hampers China's attempts at development then the use of
renewables, rather than any more polluting energy source,
becomes an essential part of the country's industrial strategy.

Similarly in Brazil, the use of biofuels that has kept the
country's emissions levels comparatively low came about
because it made economic sense rather than because of climate
concerns, as biofuels are a cheap and accessible by-product of
the sugar harvest. Brazil's use of biofuels also gives the country some energy security, as it is less dependent on imported,
and increasingly expensive, fossil fuels.

However, an overlap with other economic priorities will
not be enough to spur the development of renewables so that
that they dominate energy generation. While international
institutions such as the World Bank (see Chapter 3) need to
shift their funding priorities, governments also need to provide
dedicated support to renewables. For example, part of the
success of biofuels in Brazil is down to the support given by
the Brazilian government, which has heavily subsidised the
development of biofuels. The EREC and Greenpeace *Energy
[R]evolution* report said:

> Without political support ... renewable energy remains
> at a disadvantage, marginalised by distortions in the
> world's electricity markets created by decades of
> massive financial, political and structural support to
> conventional technologies and the failure to internalise
> environmental and social costs in price of energy.[52]

The report's recommendations included guaranteeing stable tariffs for renewable energy over a couple of decades, as has happened in Germany, where payments are guaranteed for 20 years. The price paid varies according to the relative maturity of the particular technology and reduces each year to reflect falling costs.[53] Germany is also experimenting with load management – managing the level and timing of electricity demand by giving consumers financial incentives to reduce or shut off their supply at periods of peak consumption. The Greenpeace/EREC report said:

> Many household electrical products or systems, such as refrigerators, dishwashers, washing machines, storage heaters, water pumps and air conditioning, can be managed either by temporary shut-off or by rescheduling their time of operation, thus freeing up electricity load for other uses.[54]

With the right support, renewables can contribute to a stable climate. And this contribution can begin straightaway.

## Can less be more?

This is still not enough though: the problem runs deeper than source, or style of delivery. There is also a problem with the scale of demand. The level of emissions cuts needed to stabilise the climate, combined with a lessening supply of cheap, accessible oil, makes it impossible for richer countries to carry on current consumption patterns. There has to be a limit to the growth of activities that depend on energy, making it also impossible for poorer countries to adopt similar patterns. Lester R Brown, president of the Earth Policy Institute and author of *Plan B 2.0: Rescuing a Planet Under Stress and a Civilisation in Trouble*, says:

The western economic model – the fossil fuel-based, car-centred, throwaway economy – is not going to work for China. If it does not work for China, it will not work for India, which by 2031 is projected to have a population even larger than China's. Nor will it work for the 3 billion other people in developing countries who are also dreaming the 'American dream'.[55]

Rejecting the Western economic model is a difficult thing to do. For while richer countries are seeing social isolation, high levels of debt and environmental devastation that includes the destruction of the climate on which we depend, citizens in these countries may also experience improved healthcare and a broader range of work and life opportunities. Richer countries also have some ability to adapt to a changing climate. Already, as poorer countries get richer they are adopting the transport systems, diets, leisure patterns and lifestyles typical of richer countries.[56] Changing the expectations of the 'haves' and the aspirations of the 'have-nots' will take some doing.

In richer countries, changes in the structure of economies and the pattern of demand, as well as switches in fuels, increases in energy efficiency, and shifts to less energy-intensive products have weakened a direct link between energy consumption and economic growth.[57] The structural change here includes the shift of manufacturing to other countries, where some attempts are underway to avoid the energy intensity that has traditionally characterised industrial development. These are still more 'attempts' than 'achievements', with China so far failing to meet its ambitious target of reducing the amount of energy used for each unit of gross domestic product by 20 per cent from 2006 to 2010.[58]

Such an ambition will certainly fail if increases in population and incomes create greater demand and any energy savings that are made get used up elsewhere anyway. The bottom line is that unlimited economic growth on a planet with limited resources – including limited ability of the atmosphere to cope with greenhouse gases – does not add up. If we carry on as we are we will exhaust the carrying capacity of the planet, which will support us no more.

Richer countries, where most consumption takes place – and where much of the media that portrays 'happiness' to the world is based – are the places where change in consumption patterns is most needed. A useful concept to address consumption is the idea of 'energy descent', defined by the *transitionculture.org* website as a process of:

> continual decline in net energy supporting humanity, a decline which mirrors the ascent in net energy that has taken place since the Industrial Revolution. It also refers to a future scenario in which humanity has successfully adapted to the declining net energy availability and has become more localised and self-reliant.[59]

*transitionculture.org* sees the need to transform energy use as an opportunity to develop better ways of living. Rather than wait for limits on resources to bring instability, chaos and conflict, the website suggests taking a look at:

> where we want to get to, how do we do it and what might it look like ... how [do] we design this transition in such a way that people will embrace it as a common journey, as a collective adventure, as something positive?[60]

The process it proposes for doing this is 'energy descent action planning' – planning a positive (and low-energy) vision for our communities for, say, 20 years in the future, then backcasting from then to now and producing a step-by-step timetable for managing the transition.

Towns in Ireland and England are already developing energy descent action planning, transforming how they source their food, how they travel, how they design their urban environments, how they house themselves and how they earn money; and are lobbying and working with political representatives to make these changes secure. This is a small beginning. The process would be challenging – but not impossible – for a big city, for example. However, although the process of energy descent action planning has just begun, it offers hope for a way out of the energy and resource crisis in which we have found ourselves. It also offers hope that we can find a way of living within environmental limits, in a more equitable way.

## Box 5.6: Powerdown

In his book *Powerdown: Options and Actions for a Post-Carbon World*, Richard Heinberg explores the four principal options he believes are available to industrial societies during the next few decades:

◆ *Last one standing.* The path of competition for remaining resources
   If the leadership of the United States continues with current policies, the next decades will be filled with war, economic crises and environmental catastrophe. Resource depletion and population pressure are about to catch up with us, and no one is prepared. The political elites, especially in the United States, are incapable of dealing with the situation. Their preferred 'solution' is simply to commandeer other nations' resources, using military force.

◆ *Powerdown*. The path of cooperation, conservation and sharing

The only realistic alternative to resource competition is a strategy that will require tremendous effort and economic sacrifice in order to reduce per capita resource usage in wealthy countries, develop alternative energy sources, distribute resources more equitably, and humanely but systematically reduce the size of the human population over time. The world's environmental, anti-war, anti-globalisation and human rights organisations are pushing for a mild version of this alternative, but for political reasons they tend to de-emphasise the level of effort required, and to play down the population issue.

◆ *Waiting for a magic elixir*. Wishful thinking, false hopes and denial

Most of us would like to see still another possibility – a painless transition in which market forces come to the rescue, making government intervention in the economy unnecessary. I discuss why this rosy hope is extremely unrealistic, and serves primarily as a distraction from the hard work that will be required in order to avert violent competition and catastrophic collapse.

◆ *Building lifeboats*. The path of community solidarity and preservation

This fourth and final option begins with the assumption that industrial civilisation cannot be salvaged in anything like its present form, and that we are even now living through the early stages of disintegration. If this is so, it makes sense for at least some of us to devote our energies toward preserving the most worthwhile cultural achievements of the past few centuries.

Heinberg believes that:

> attempting to maintain business as usual during the coming decades will merely ensure catastrophic collapse. However, we can preserve the best of what we have achieved, while at the same time easing our way as peacefully and equitably as possible back down the steep ramp of increasing scale and complexity our society has been climbing for the past couple of centuries. These are the options we face, and the sooner we acknowledge that this is the case and choose wisely, the better off we and our descendants will be.

Source: Richard Heinberg, *Powerdown: Options and Actions for a Post-Carbon World*, New Society Publishers, 2004, pp. 14–15.

# Conclusion:
# Requiem for a lost season

'Of what are you thinking?' asked Avoa.

'I am thinking of the moment something dies and how we instinctively know it. And of how we try not to know what we know because we do not yet understand how we are to negotiate change'.

Alice Walker, *Now is the Time to Open Your Heart*[1]

## Hot when it should be cold, dry when it should be wet

Apart from a few brief spells of cold, in England the winter of 2006/7 hardly happened. On Christmas Day – surrounded by Christmas cards featuring robins on snow-covered spade handles, robins on snow-covered branches, snowmen, snow-covered church spires, children skipping gaily in the snow – we lit a log fire in the sitting room hearth. This seemed like a cosy thing to do at this time of year known as the 'bleak midwinter'. According to the Christmas carol of that name, the earth should have been standing as hard as iron and water should have been like a stone. Apparently a frosty wind should have been moaning too, but all we got as we opened the windows to let out the heat was a nice fresh breeze. In my neighbours' garden the golden trumpets of the daffodils, which herald spring and the coming of warmer months, were fully open a week later. The buds on the camellia bush that usually signal it's time to put away warm hats and gloves were on full view before my winter woollens had even had an airing.

During this unusually warm winter the writer and campaigner George Marshall, blogging at the website *climate denial.org*, said:

> The prognosis is clear. Assuming that there is no collapse of the Gulf Stream, within one generation low land Western Europe will become permanently devoid of snow. We're virtually there already in southern England. The number of days of snow has fallen by two thirds during my lifetime. The heaviest snowfall my children have experienced in their five years of life is one inch three years ago. Since then there has been hardly enough to speckle the path.[2]

By the time my daughter, currently less than a year old, is big enough to work out what a Christmas card is, she may well be wondering why it pictures this snow stuff when all it does at Christmas is rain. And rain. And rain.

If that's all she has to worry about then she'll be lucky. At the moment we have enough resources to cope with the changes in the climate that our family faces. If our garden doesn't produce enough food because the unseasonably warm weather means that frogspawn is laid too early to survive, resulting in no frogs to eat pests on our vegetables, then we can just go to the shop. If the shop has problems sourcing one type of vegetable they'll fill the shelf space with another. We are lucky to be insulated in such a way. In other parts of the world a lost season has a different significance.

In Russia, the lack of December snow in a warmer than usual winter led TV channels to predict an outbreak of depression.[3] This was not because of concern over natural patterns being so strangely out of tune: it was because the snow that usually blankets the ground at that time of year reflects the little sunlight there is between the long nights, triggering mood-enhancing chemicals in the brain.

In the United States, in 2006, representatives from more than 50 tribes held the Tribal Lands Climate Conference, a two-day conference on climate change.[4] Coming from Alaska in the north to Arizona in the south, each tribe said it was experiencing changes in precipitation, temperature and wildlife that appeared to be brought on by climate change. These alterations were threatening land, health and culture. 'We basically have two seasons now,' said Robert Gomez, director of the environmental office of the Taos Pueblo reservation in northern New Mexico. 'Hot and dry, and cold and dry.'[5] Climate change has a particular effect on tribes that are strongly connected to a particular location or reservation: 'As our species migrate off, we don't have the legal right to follow them,' said Terry Williams, fisheries and natural resources commissioner for the Tulalip Tribes.[6] Williams warned that, if nothing is done: 'within the next 20 to 25 years, our culture will be terminated, because the necessary species will be gone.'

Drier than usual seasons also pose a real problem for Australia's southeastern region. Mean temperatures in Australia have increased faster than the global average since 1910.[7] However, the severe drought affecting the country's southeast has been masked in national rainfall figures by unusually wet weather in the north and west. This trend to more marked droughts is affecting the country's most important agriculture regions. Australia is usually one of the world's top three grain exporters, and a sharp reduction in its expected wheat crop has already pushed up global prices. A report by the Climate Action Network of Australia predicted that reduced rainfall as a result of climate change could lead

to a 15 per cent drop in grass productivity,[8] which could lead
to one problem after another:

> [The drop in grass productivity], in turn, could lead to
> reductions in the average weight of cattle by 12 per cent,
> significantly reducing beef supply. Under such conditions,
> dairy cows are projected to produce 30 per cent less milk,
> and new pests are likely to spread in fruit-growing areas.
> Additionally, such conditions are projected to lead to 10
> per cent less water for drinking. Based on model projec-
> tions of coming change conditions such as these could
> occur in several food producing regions around the world
> at the same time within the next 15–30 years.[9]

A 2003 report on the implications for US national security of
an abrupt climate change scenario suggested the situation
outlined by Australia's Climate Action Network '[challenged]
the notion that society's ability to adapt will make climate
change manageable'.[10]

## Anomalies

Is climate change to blame for the disappearing seasons? Some
people still argue that naturally occurring weather patterns
may be behind it all. One such weather pattern is the El Niño
climate event, which involves changes in ocean currents and
atmospheric circulation. While climate events such as El Niño
do have an impact on the climate system, the scale of current
changes suggests there is something more at work.
Commenting on unseasonably warm temperatures in the
United States, *Real Climate*, a website where climate scientists
comment on climate science, said:

> while El Niño typically does perturb the winter Northern

Hemisphere jet stream in a way that favors anomalous warmth over much of the northern half of the United States, the typical amplitude of the warming is about 1°C. The current anomaly is roughly five times as large as this. One therefore cannot sensibly argue that the current US winter temperature anomalies are attributed entirely to the current moderate El Niño event.[11]

The *Real Climate* website was hesitant to say that such unusual warming is necessarily due to climate change. It said: 'one cannot attribute a specific meteorological event, an anomalous season ... to climate change.'[12] However, the website did concede that: 'one can argue that the pattern of anomalous winter warmth seen last year, and so far this year, is in the direction of what the [climate] models predict.'[13] The website also pointed out that even if an unusual weather event could be put down to El Niño, this would not mean that climate change did not play a role.[14] Many scientists argue that El Niño itself is being influenced by climate change, although there is a debate over the extent to which this is going on.

The writer Fred Pearce says that all views on climate change should be included in the debate as this would ensure that predictions are not narrowed down to just the ones that everyone agrees with.[15] Pearce's interest lies at the other end of the debate – rather than believing that global warming is having little impact on weather systems, he believes that it may have an even greater impact than predicted, with abrupt and dramatic events more likely than previously thought. The IPCC reports that underpin international climate discussions rarely consider these extremes. The reports are drawn up on the basis of consensus, ruling out conclusions that are exceptionally dramatic, and tending towards a conclusion that climate change will be a gradual process, rather than a scenario of unpredictable twists and turns.

## Feasible targets

One conclusion that does often find agreement is the difficulty of stabilising (all) greenhouse gas emissions at 450ppm – a level that is necessary to avoid dangerous climate change. *The Economics of Climate Change: The Stern Review* warned:

> It is instructive that cost modelling exercises rarely consider stabilisation below 500ppm $CO_2e$ (carbon dioxide equivalence). ... [Some models] simply cannot find a way of achieving 450ppm $CO_2e$. ... 450ppm $CO_2e$ would in addition require very large and early reductions of emissions from transport, for which technologies are further away from deployment.[16]

The *Stern Review* said that a more economically feasible target for greenhouse gas concentrations was 550ppm, a figure that has a 78 to 99 per cent chance of exceeding 2°C.[17] Although the *Stern Review* was clear about the dangers of a temperature rise greater than 2°C, its conclusions were that such dangers were inevitable. It seems that many people, the poor in particular, are condemned to increasingly dangerous climate change impacts. On hearing Stern's conclusions, Christian Aid pointed out that the Review's figure for what was economically feasible was incompatible with meeting the needs of the world's poor. Christian Aid said:

> We are concerned that the Stern Report has dismissed stabilising global emissions at a level of $CO_2$ and other equivalent greenhouse gases of 450 parts per million as too expensive. ... In reality poor people are already struggling to cope with existing climate change as a result of an atmosphere polluted with 430ppm of $CO_2e$.

At Stern's levels (550ppm of $CO_2$e), large parts of the developing world would still be exposed to a much greater risk of disaster and misery.[18]

## Political realism

Bringing emissions down to a level that does not mean catastrophe will require actions that are not only, in Stern's view, economically unrealistic. The necessary actions are also, in the current climate, politically unrealistic. Between 2000 and 2005 the rate of increase in emissions from burning fossil fuels was four times that between 1990 and 2000.[19] And, despite Chapter 5's optimism that a less polluting form of development is possible, many countries are counting on higher oil consumption in the decades ahead. Car assembly plants, roads, car parks and suburban housing developments are being built as though climate change is a figment of the imagination.[20] New aeroplanes are being made with the expectation that air travel and freight will expand indefinitely.[21] China's civil aviation fleet is due to increase from 777 planes in 2003 to over 2800 in 2023.[22]

No industrialised country has committed to the level of cuts needed to have a genuine chance of staying within the limit of a temperature rise of no more than 2°C above the average global temperature in pre-industrial times. This is not because they don't recognise the importance of this temperature threshold. The EU, for example, has committed to a limit on temperature rise of no more than 2°C. One of the vehicles for the EU to reduce its own greenhouse gas emissions is the EU Emissions Trading Scheme (ETS). The first phase of the EU ETS ran from 2005 to 2007. It gave a limited number of emissions allowances to heavy energy-using industries – electricity generators; oil refineries; iron, steel and minerals industries; paper, pulp and board manufacturers. However, as the end of the first phase of the EU ETS approached, there was little evidence that businesses taking part in the scheme had shifted to cleaner fuels. In fact things were going the other way – in 2006 the EU burned approximately 10 million tonnes more coal than it had done in 2005.[23] The same year saw the UK import the largest amount of coal in its history.[24] A 2004 briefing paper from the environmental consultants Enviros had predicted that the ETS's first phase would be ineffective, suggesting that the allocation of emissions allowances would lead to European industry emitting 5–11 per cent more carbon dioxide than in 2000.[25] An over-generous allocation of emissions allowances did indeed turn out to be a problem, and the price of carbon plunged to half its value in one week in April 2006 as some countries began to report their actual level of emissions. Reflecting on the drop in carbon price, the UK's *Financial Times* said:

> there is a strong suspicion that EU governments, of which at least 15 are on track to exceed their eventual Kyoto targets, are being too generous in awarding permits to their industries rather than the latter being unexpectedly successful in cutting pollution.[26]

Although a political framework for effective emissions cuts seems politically unrealistic, it is increasingly essential. The report from the 2005 conference on avoiding dangerous climate change said: 'For all stabilisation strategies, the biggest problem does not seem to be the technologies or the costs, but overcoming the many political, social and behavioural barriers to implementing mitigation options.'[27]

In England, where I live, all of the three main political parties fall into the category of being a barrier 'to implementing mitigation options'. None of them base their policies in an understanding that growth cannot continue indefinitely. None of them have effective policies for low-emission transport systems, for example, or low-emission power generation. All of them act as though strategies that could reduce emissions to the level needed don't actually exist. All of them often talk, but fail to act, bold.

## Political context

If politicians are to be bold without committing political suicide, we have to create a wider context that demands action on climate change. This is as much our responsibility as it is theirs – making the demand for action so irresistible that they can ignore it no longer. We can do this through pressuring for change, and through highlighting examples where change is already underway. One such example is the 2007 agreement by the European Union to cut greenhouse gas emissions by 20 per cent by 2020 compared with 1990 levels. The EU also offered a carrot to the rest of the world by suggesting it would cut emissions by 10 per cent more if others followed suit.[28] While this needs to be backed up with targets and planning – and needs bigger cuts to follow – it is a step in the right direction, especially as it includes an offer to talk about further cuts. The EU must not be allowed to back away from this. This kind of pre-commitment offer, which does not become binding unless reciprocal offers are made, is a step towards breaking the impasse over who moves first in international climate negotiations. It is an important step to take for countries such as the UK, where the share of global emissions is low but political influence on global issues is high. This is not the only effect that the UK could have. It could also influence action on climate change by showing that it is possible to shift from a carbon-intensive pattern of industrial development to one that has a lower impact. The UK is rich enough to develop and try out low-carbon technologies and energy systems, making these more affordable for other countries to use. It also has the resources and the influence to press for greater support for adaptation.

Another positive example to promote and learn from is Sweden's commitment to replace fossil fuels with renewables over the next 15 years. Mona Sahlin, Sweden's minister of sustainable development said:

> The aim is to break dependence on fossil fuels by 2020. By then no home will need oil for heating. By then no motorist will be obliged to use petrol as the sole option available. By then there will always be better alternatives to oil.[29]

Sweden has a head start as its current dependence on oil is small. Oil is mostly just used for transport. Also, it already has a high level of renewable use: in 2003, 26 per cent of all the energy consumed in Sweden already came from renewable sources, compared with the EU average of 6 per cent.[30]

## Individual actions

In richer countries, while looking outwards and putting pressure on political representatives to take action on climate change, there are many useful changes we can make by looking inwards and making changes in our own lives. In isolation, individual behaviour may not make much of a difference. In an essay titled 'The Irony of Environmentalism: The Ecological Futility but Political Necessity of Lifestyle Change', Paul Wapner and John Willoughby warned that there are limits to the impact of individual change. They said that 'individual action within the current world economy will not reduce overall thoroughput, but will simply change where the engines of consumption operate.' The value of individual action, according to Wapner and Willoughby, was more about the following:

> if we are ever to create an ecologically sound world in which social structures and individuals operate in ways that enhance environmental well-being it will be based on, and inspired by, the model of action individual environmentalists undertake in their personal lives.[31]

Wapner and Willoughby saw the benefits of individual choice in environmental politics 'more in moral terms and for the good of political agency rather than as a direct causal influence on ecological conditions'. Individual action is most valuable when it contributes to changing the wider political context of concern over the issue. This could mean supporting campaigns against aviation expansion, for example, while also personally cutting down on flying. (Aviation makes a massive contribution to greenhouse gas emissions. It would account for 50 per cent of all UK emissions if the UK aimed to stabilise greenhouse gases at 550ppm.[32] It would exceed the UK's entire carbon budget by 2050 if the UK aimed to stabilise greenhouse gases at 450ppm.[33]) It could mean sourcing household energy from a company that uses renewable sources while also developing a community-wide energy supply company to use local wind, water, solar or other renewable resources. It could mean putting time aside to attend a climate event and adding your energy to a wider movement for action on climate change. Writing about the importance of a social movement on climate change, George Marshall said:

> People will not accept the reality of the problem unless they see that others are engaging in activities that reflect its seriousness. This means they need to be confronted by emotionally charged activities; debate, protest, and

meaningful visible alternatives. Simply asking people to change their light bulbs, plant a tree, or send in a donation, however desirable in themselves, will not build a social movement. The activities are not proportionate to the level of threat and will persuade no one.[34]

## Anyone but us

The above examples illustrate that it is possible to make wider ripples from change that starts at home. But the motivation to take such steps has to begin with an understanding that we, particularly those of us who live in the world's richer countries, are part of the problem. As this book has tried to show, the bulk of greenhouse gas emissions in the next few decades will come from countries that have, historically, been low emitters. Even though these countries have contributed so little towards actually creating the problem (and are now producing high emissions largely to meet rich world demand) they are likely to be blamed for the climate crisis. 'It's the Chinese, isn't it?' is a not uncommon response when talking to people here in the UK about climate change. No, it's not just the Chinese: the problem is not just 'out there' and the fault of someone else – the Chinese, the Indians, the Russians (again). It's not even just the fault of the United States. As Chapter 4 said, emissions per person are due to remain highest in the richer countries. These are the places where people's lifestyles are so caught up with high use of fossil fuels that the problem has to lie, at least to some extent, with each and every one of us that lives in such a high-emitting way.

This can be a difficult fact to accept. It means accepting that we may have messed up the environment for our own and future generations. It means accepting that, just as we have the power to cause damage to the environment, we also have the power to repair it. We have the power to limit the impact of the changes that are already locked in, and we have the power to limit the degree of change that will take place in the future. And that is quite a responsibility.

## The great turning

The writer and activist Joanna Macy talks about the time in which we are living as 'the great turning'. This is a positive spin on the current situation of ecological disaster: the great turning is the time in which a self-destructive industrial-growth society turns into a life-sustaining society. In their book *Coming Back to Life: Practices to Reconnect Our Lives, Our World*, Macy and Molly Young Brown say:

> The most remarkable feature of this historical moment on Earth is not that we are on our way to destroying our world – we've actually been on the way for quite a while. It is that we are beginning to wake up, as from a millennia-long sleep, to a whole new relationship to our world, to ourselves and each other.[35]

They see this moment as hugely important:

> When people of the future look back at this historical moment, they will see, perhaps more clearly than we can now, how revolutionary it is. ... They will see it as epochal. While the agricultural revolution took centuries, and the industrial revolution took generations, this ecological revolution has to happen within a matter of a few years. It also has to be more comprehensive – involving not only the political economy, but the habits and values that foster it.[36]

Macy and Brown believe that we have numbed our hearts and

minds in order to cope with the distress that we see around us every day – the environmental destruction, the effects of poverty, the wars. We blame others for the problem because we want someone else to be responsible for it, whatever it is. We hope that our political leaders will do something. We distract ourselves with shopping. We avoid the information when it all gets too much. Macy and Brown aim to help people become 'enlivened and motivated to play their part in creating a sustainable civilisation'.[37] To move beyond the numbing and get to that point, they say that we must first recognise that what we are feeling is pain. A participant in one of Macy's workshops said:

> My name is Mark. I work on contract to the Navy, consulting on weapons systems. This week my little boy was sorting books for a school sale, and asked if he should keep some of his favourites to pass on to his own children. I could hardly answer because I realised that I doubted he would live that long.[38]

Only when you've acknowledged this pain can you step forward and act. Macy and Brown quote from Jack Kornfield:

> Strength of heart comes from knowing that the pain that we each must bear is part of the greater pain shared by all that lives. It is not just 'our' pain but the pain, and realising this awakens our universal compassion.[39]

Climate change is already a cause of pain. Those exposed to its impacts are feeling pain. The knowledge that we are inevitably heading for a warmer world and things are going to get worse before they get better is painful. The sight of horse-trading in international negotiations while emissions shoot up is painful. The knowledge that my child will grow up in an increasingly unstable world is painful.

I think that Macy and Brown bring an important element to the climate debate, even though their analysis and style of speaking emotions out loud certainly doesn't suit everyone. Their work is rare in acknowledging that the level of instability in the climate system is not just a physical fact: it is also a deeply upsetting phenomenon. It can, and will, ruin lives. It is taking place because our highly sophisticated, technologically advanced society, which has split the atom and put a man on the moon, is now destroying its own conditions for life. Coming to terms with this on an emotional level, in whatever way suits each individual, may be an important hurdle to cross before we get to the point where we take on our ability to influence change – to exercise political agency. For, like so many others, Macy and Brown say that if we fail to make changes (in this case, to stabilise the climate) 'it will not be for lack of technology or relevant data so much as for lack of political will'.[40]

# Resources

## Bibliography

The following books offer useful background reading for a more in-depth coverage of the issues presented in this guide.

Brown, Lester R., *Plan B 2.0: Rescuing a Planet Under Stress and a Civilisation in Trouble,* Norton, 2006.
The global scale and growing complexity of issues facing our world have no precedent. Brown highlights how, in ignoring nature's deadlines for dealing with these issues, we risk the disruption of economic progress.

Davidson, J. and Myers, D., *No Time to Waste: Poverty and the Global Environment*, Oxfam, Oxford, 1992.
This book outlines the connections between the issues.

Development Dialogue no. 48, *Carbon Trading: A Critical Conversation on Climate Change, Privatisation and Power*, Dag Hammarskjold Foundation, 2006.
This special edition of the *Development Dialogue* journal discusses the role of carbon trading within development.

Heinberg, Richard, *Powerdown: Options and Actions for a Post-Carbon World*, Clairview Books, 2004.
*Powerdown* gives an overview of the likely impacts of oil and natural gas depletion and outlines options for industrial societies during the next decades.

Heinberg, Richard, *Party's Over: Oil, War and the Fate of Industrial Societies*, Clairview Books, 2005.
Heinberg places the imminent energy transition in historical

context and describes the likely impacts of oil depletion, and the energy alternatives.

Kolbert, Elizabeth, *Field Notes from a Catastrophe: Man, Nature, and Climate Change*, Bloomsbury, 2006.
Kolbert travels to the Arctic, explains the science, draws parallels to ancient civilisations, unpacks the politics, and presents the personal tales of those who are being affected most.

Lynas, Mark, *High Tide: News from a Warming World*, HarperCollins, 2005.
Lynas travels around the world to show the impacts of global warming already being felt in people's lives.

Lynas, Mark, *Six Degrees: Our Future on a Hotter Planet*, Fourth Estate, 2007.
Lynas presents an account of the future of our earth and our civilisation if current rates of global warming persist.

Macy, Joanna R. and Young Brown, Molly, *Coming Back to Life: Practices to Reconnect Our Lives, Our World*, New Society Publishers, 1998.
A look at the angst of our era – pain, fear, guilt and inaction – and a pointer to the way out of apathy.

McKibben, Bill, *The End of Nature, Bloomsbury*, 2003.
McKibben explains the implications of the destruction wrought on our planet, including the end of nature as something independent of, larger than, and uncontrolled by man.

Meyer, Aubrey, *Contraction and Convergence: The Global Solution to Climate Change (Schumacher Briefings)*, Green Books, 2000.
This briefing explains the global policy framework of 'Contraction and Convergence'.

Monbiot, George, *Heat: How to Stop the Planet Burning*, Allen Lane, 2006.

Monbiot considers what must be done to reduce greenhouse gas emissions to a safe level.

Pearce, Fred, *The Last Generation: How Nature Will Take Her Revenge for Climate Change*, Eden Project Books, 2006.

Pearce visits the places where dramatic climate change may start and uncovers the first signs that nature's revenge is already under way.

Hans-Joachim Schellnhuber, Wolfgang Cramer, Nebojsa Nakiccnovic, Tom Wigley and Gary Wynn Yohe (eds.), *Avoiding Dangerous Climate Change*, Cambridge University Press, 2006.

Download for free at http://www.defra.gov.uk/environment/climatechange/internat/pdf/avoid-dangercc.pdf.

In 2005 the UK Government hosted the Avoiding Dangerous Climate Change conference to look at the scientific issues associated with climate change. This volume presents findings from the leading international scientists who attended the conference.

Simms, Andrew, *Ecological Debt: The Health of the Planet and the Wealth of Nations*, Pluto Press, 2005.

This book illustrates the harmful consequences of unsustainable consumption patterns and shows how to make changes to help to preserve the balance of the environment.

## Publications and web resources

As well as the books listed above, the following publications and web-based resources can help with further research. They relate to the themes of each chapter and are grouped as follows: adaptations and development issues, carbon trading, international finance, international programmes, and energy issues. Publications or web resources that provide extensive coverage of more than one of the themes are listed at the end under 'general background'.

## Adaptations and development issues

W. Neil Adger, Nick Brooks, Graham Bentham, Maureen Agnew and Siri Eriksen, *New Indicators of Vulnerability and Adaptive Capacity*, Tyndall Centre for Climate Change Research Technical Report 7, 2004.

Christian Aid, *The Climate of Poverty: Facts, Fears and Hope*, 2006.

Saleemul Huq, *Adaptation to Climate Change: A Paper for the International Climate Change Taskforce*, Institute for Public Policy Research, 2005.

John Magrath, *Glacier Melt: Why it Matters for Poor People*, Oxfam, 2004.

Oxfam, *Adapting to Climate Change: What's Needed in Poor Countries and Who Should Pay*, Oxfam, 2007.

David Thomas, Henny Osbahr, Chasca Twyman, Neil Adger and Bruce Hewitson, *ADAPTIVE: Adaptations to Climate Change Amongst Natural Resource-dependent Societies in the Developing World: Across the Southern African Climate Gradient*, Tyndall Centre for Climate Change Research Technical Report No. 35, 2005.

## Carbon trading

Jessica Ayres, Maryanne Grieg-Gran, Lizzie Harris and Saleemul Huq, *Expanding the Development Benefits from Carbon Offsets*, International Institute for Environment and Development, 2006.

Heidi Bachram, 'Climate Fraud and Carbon Colonialism: The New Trade in Greenhouse Gases', *Capitalism Nature Socialism*, December 2004.

Graham Erion, *Low Hanging Fruit Always Rots First: Observations from South Africa's Crony Carbon Market*, Center for Civil Society, University of KwaZulu-Natal, South Africa, 2005.

Larry Lohmann, 'Marketing and Making Carbon Dumps: Commodification, Calculation and Counterfactuals in Climate Change Mitigation', *Science as Culture*, Vol. 14, September 2005.

## International finance

Elizabeth Bast and David Waskow, *Power Failure: How the World Bank is Failing to Adequately Finance Renewable Energy for Development*, Friends of the Earth – United States, October 2005.

CDM Watch, *The World Bank and the Carbon Market: Rhetoric and Reality*, April 2005.

Jon Sohn, Smita Nakhooda and Kevin Baumert, *Mainstreaming Climate Change Considerations at the Multilateral Development Banks*, World Resources Institute, July 2005.

Jim Vallette and Steve Kretzmann, *The Energy Tug of War: The Winners and Losers of World Bank Fossil Fuel Finance*, Sustainable Energy & Economy Network, April 2004.

World Bank, *Striking a Better Balance: The Extractive Industries Review*, December 2003.

World Bank, *Clean Energy and Development: Towards an Investment Framework*, April 2006.

World Bank, *State and Trends of the Carbon Market 2006*, May 2006.

World Bank Carbon Finance Unit, *The Role of the World Bank in Carbon Finance: An Approach for Further Engagement*, January 2006.

## International programmes

T. Athanasiou, S. Kartha and P. Baer, *Greenhouse Development Rights: An Approach to the Global Climate*

Regime that Takes Climate Protection Seriously While Also Preserving the Right to Human Development, EcoEquity and Christian Aid, 2006.

Paul Baer with Michael Mastrandrea, High Stakes: Designing Emissions Pathways to Reduce the Risk of Dangerous Climate Change, Institute for Public Policy Research, 2006.

Aaron Cosbey, Warren Bell and John Drexhage, Which Way Forward? Issues in Developing an Effective Climate Regime after 2012, International Institute for Sustainable Development, 2005.

Kate Hampton, Catalysing Commitment on Climate Change, Institute for Public Policy Research, 2005.

## Energy issues

K. Anderson, S. Shackley, S. Mander and A. Bows, Decarbonising the UK: Energy for a Climate Conscious Future, Tyndall Centre for Climate Change Research Technical Report No. 33, 2005.

Helen Buckland, The Oil for Ape Scandal: How Palm Oil is Threatening Orang-Utan Survival, Friends of the Earth, Ape Alliance, Borneo Orangutan Survival Foundation, Orangutan Foundation (UK) and Sumatran Orangutan Society, September 2005.

Robert L. Hirsch, Roger Bezdek and Robert Wendling, Peaking of World Oil Production: Impacts, Mitigation, and Risk Management, US Department of Energy, February 2005.

Rob Hopkins, Energy Descent Pathways: Evaluating Potential Responses to Peak Oil, MSc Dissertation for the University of Plymouth, 2006.

Peter Schwartz and Doug Randall, An Abrupt Climate Change Scenario and Its Implications for United States National Security, Global Business Network, October 2003.

Andrew Simms, Julian Oram and Petra Kjell, *The Price of Power: Poverty, Climate Change, the Coming Energy Crisis and the Renewable Revolution*, New Economics Foundation, 2004.

Sven Teske, Arthouros Zervos and Oliver Schäfer, *Energy [R]evolution: A Sustainable World Energy Outlook*, Greenpeace and the European Renewable Energy Council, January 2007.

Transition Culture: a website about solutions to peak oil using managed energy descent and permaculture – www.transition culture.org.

United Nations Environment Programme, *Changing Climates: The Role of Renewable Energy in a Carbon-Constrained World*, pre-publication draft, Renewable Energy Policy Network (REN21), December 2005.

US Department of Energy, *International Energy Outlook 2005*, July 2005.

Worldwatch Institute, *Energy for Development: The Potential Role of Renewable Energy in Meeting the Millennium Development Goals*, Renewable Energy Policy Network (REN21), 2005.

Worldwatch Institute, *Renewables 2005 Global Status Report*, Renewable Energy Policy Network (REN21), 2005.

### General background

Carbon Trade Watch, *Hoodwinked in the Hothouse: The G8, Climate Change and Free-market Environmentalism*, Transnational Institute briefing series No 2005/3, June 2005.

Climate Denial: a blog exploring our deep and profound denial of climate change – www.climatedenial.org.

DeSmogBlog: clears away the PR pollution clouding the science on climate change – www.desmogblog.com.

Global Issues: outlines global issues and shows how these are interrelated – www.globalissues.org.

RealClimate: a commentary site on climate science by working climate scientists for the interested public and journalists – www.realclimate.org.

Mark Lynas's blog: blog from the author of *High Tide: News from a Warming World* and *Six Degrees: Our Future on a Hotter Planet* – www.marklynas.org.

Andrew Simms, John Magrath and Hannah Reid with contributions from the Working Group on Climate Change and Development, *Up in Smoke? Threats from, and Responses to, the Impact of Global Warming on Human Development*, New Economics Foundation, 2004.

Andrew Simms with support from Hannah Reid, *Africa: Up in smoke? The Second Report from the Working Group on Climate Change and Development*, New Economics Foundation, June 2005.

Nicholas Stern, *The Economics of Climate Change: The Stern Review*, Cambridge University Press, 2007.

Tiempo: an electronic information service covering global warming, climate change, sea level rise and related issues – www.tiempocyberclimate.org.

United Nations, *Convention on Biological Diversity*, 5 June 1992.

United Nations, *United Nations Framework Convention on Climate Change*, 1992.

United Nations, *Millennium Ecosystem Assessment*, 2005.

UNFCCC: background and latest news on the United Nations Framework Convention on Climate Change – unfccc.int.

Paul Wapner and John Willoughby, 'The Irony of Environmentalism: The Ecological Futility but Political

Necessity of Lifestyle Change', *Ethics & International Affairs*, Vol. 19, No. 3, November 2005.

## Organisations

**Camp for Climate Action**
A UK-based initiative to explore grassroots solutions to climate change through workshops, skill-sharing, education, debate, entertainment and direct action.
www.climatecamp.org.uk
Tel: +44 (0) 777 286 1099
Email: info@climatecamp.org.uk
Post: Cornerstone Resource Centre, 16 Sholebroke Avenue, Leeds LS7 3HB, UK

**Carbon Trade Watch**
A meeting point for researchers, campaigners and communities opposing the negative impacts of pollution trading. It monitors the impact of pollution trading upon environmental, social and economic justice.
www.carbontradewatch.org
Amsterdam office:
Tel: + 31 20 662 66 08
Post: Transnational Institute, PO Box 14656, 1001 LD Amsterdam, Netherlands
CTW in Spain:
Tel/fax: +34 985 493696
Email: tamra@carbontradewatch.org
Post: TNI/CTW, ESCANDA, Tamra Gilbertson, La Casona, Ronzon, 33637 Lena, Asturias, Spain
CTW in the UK:
Tel: +44 (0) 207 700 7971
Email: kevin@carbontradewatch.org

**Christian Aid's climate campaign**
Christian Aid is an agency of the churches in the UK and
Ireland. It works for a world transformed by an end to poverty
and campaigns to change the rules that keep people poor.
Christian Aid's climate campaign aims to get the UK to cut its
carbon emissions by 5 per cent a year.
http://www.christian-aid.org.uk/climatechange/index.htm
Tel: +44 (0) 20 7620 4444
Fax: +44 (0) 20 7620 0719
Email: info@christian-aid.org
Post (head office): 35 Lower Marsh, Waterloo,
London SE1 7RL, UK

**Climate Action Network**
A worldwide network of over 365 non-governmental organisations working to promote government and individual action to
limit human-induced climate change to ecologically sustainable
levels.
www.climatenetwork.org
Tel: +1 202 609 9846
Fax: +1 202 536 5503
Email: jcoven@climatenetwork.org
Post: Climate Action Network – International Secretariat,
1326 14th St. NW, Washington, DC 20005, USA

**Climate Outreach and Information Network (COIN)**
Through education and innovative approaches to learning,
COIN aims to achieve permanent reductions in household
greenhouse gas emissions, reducing them to levels which can
be sustained, and which result in no further degradation of
ecological systems and human livelihood.
www.coinet.org.uk
Tel: +44 (0)1865 727911

Email: via the website
Post: 16B Cherwell Street, Oxford OX4 1BG, UK

**Corner House (The)**
Carries out analyses, research and advocacy with the aim of linking issues, stimulating informed discussion and strategic thought on critical environmental and social concerns, and encouraging broad alliances to tackle them.
www.thecornerhouse.org.uk
Tel: 0845 330 7928 (UK only, local rate)
Tel: +44 (0) 1258 473 795 (International)
Fax: +44 (0) 1258 473 748
Email: enquiries@thecornerhouse.org.uk
Post: Station Road, Sturminster Newton, Dorset DT10 1YJ, UK

**Down to Earth**
News, features, articles, perspectives and research papers on science, environment, development, poverty and sustainability issues affecting India and the world.
www.downtoearth.org.in
Tel: +91 11 2995 5124, 2995 6110, 2995 6394
Fax: +91 11 2995 5879
E-mail: downtoearth@downtoearth.org.in
Post: Society for Environmental Communications, 41, Tughlakabad Institutional Area, New Delhi, India – 110062

**EcoEquity**
A research and advocacy organisation dedicated to the promotion of a just and adequate solution to the climate crisis.
www.ecoequity.org
(web address only)

**Global Commons Institute**
An independent group concerned with the protection of the global commons.
www.gci.org.uk
Email: aubrey@gci.org.uk

**Global Cool**
Global Cool believes that the solution to tackling climate change lies within the power of the individual. It aims to become a one-stop-shop for a cleaner, more energy efficient life.
www.global-cool.com
Email: questions@global-cool.com

**International Institute for Environment and Development**
An international policy research institute and non-governmental body working for more sustainable and equitable global development.
www.iied.org
Tel: +44 (0) 20 7388 2117
Fax: +44 (0)20 7388 2826
email: info@iied.org
Post (head office): 3 Endsleigh Street, London WC1H 0DD, UK

**Intergovernmental Panel on Climate Change (IPCC)**
Assesses the scientific, technical and socio-economic information relevant to understanding the scientific basis of risk of human-induced climate change, its potential impacts and options for adaptation and mitigation.
www.ipcc.ch
Tel: +41 22 730 8208/84
Fax: +41 22 730 8025/13
E-mail: IPCC-Sec@wmo.int

Post: IPCC Secretariat, c/o World Meteorological Organization, 7bis avenue de la Paix, C.P. 2300, CH-1211 Geneva 2, Switzerland

**nef (the new economics foundation)**
An independent think-and-do tank that inspires and demonstrates real economic well-being.
www.neweconomics.org
Tel: +44 (0) 20 7820 6300
Fax: +44 (0) 20 7820 6301
Email: info@neweconomics.org
Post: 3 Jonathan Street, London SE11 5NH, UK

**Oxfam's Climate Change Campaign**
Resources and suggestions for action on climate change from Oxfam.
www.oxfam.org.uk/climatechange
Tel: +44 (0) 870 333 2700
Email: via the website
Post: Oxfam Supporter Relations, Oxfam House, John Smith Drive, Cowley, Oxford OX4 2JY, UK

**Practical Action**
Formerly know as ITDG (the Intermediate Technology Development Group), Practical Action aims to demonstrate and advocate the sustainable use of technology to reduce poverty in developing countries.
www.practicalaction.org
Tel: +44 (0) 1926 634400
Fax: +44 (0) 1926 634401
Email: practicalaction@practicalaction.org.uk
Post (head office): The Schumacher Centre for Technology & Development, Bourton on Dunsmore, Rugby CV23 9QZ, UK

**Sinkswatch**
An initiative to track and scrutinise carbon sink projects.
www.sinkswatch.org
Tel: +44 (0) 1608 652 895 or +44 (0) 7931 576538
Fax: +44 (0) 1608 652 878
Email: jutta@fern.org
Post: Jutta Kill, 1c Fosseway Business Centre, Stratford Road,
Moreton-in-Marsh, Gloucestershire GL56 9NQ, UK

**Stop Climate Chaos**
Aims to build a massive coalition that will create an irresistible
public mandate for political action to stop human-induced
climate change. Includes the I Count campaign –
www.icount.org.uk.
www.stopclimatechaos.org
Tel: +44 (0) 20 7729 8732
Email: info@stopclimatechaos.org
Post: 2 Chapel Place, London EC2A 3DQ, UK

**Sustainable Energy & Economy Network (SEEN)**
Works in partnership with citizens' groups on environment,
human rights and development issues with a particular focus
on energy, climate change, environmental justice, gender
equity and economic issues, particularly as these play out in
North/South relations.
www.seen.org
Daphne Wysham, Co-Director
Tel: 202 234 9382, X208
Fax: 202 387 7915
Post: Institute for Policy Studies, 1112 16th St. NW, Suite 600,
Washington, DC 20036, USA
Jim Vallette, Research Director
Tel: 1 207 244 3106
Fax: 1 561 431 0139

E-mail: jvallette@seen.org
Post: 400 Seawall Road, Southwest Harbor, Maine 04679, USA
Nadia Martinez, Co-Director
Tel: 202 234 9382, X209
Fax: 202 387 7915
E-mail: nmartinez@seen.org
Post: Institute for Policy Studies, 1112 16th St. NW, Suite 600, Washington, DC 20036, USA

## Tearfund climate change campaign

Tearfund is a Christian relief and development charity. It campaigns on climate change and disasters because these are issues that severely affect the lives of the world's poorest people.
http://www.tearfund.org/Campaigning/Climate+change+and+disasters/
Tel: +44 (0) 20 8977 9144
Fax: +44 (0) 20 8943 3594
Email: enquiry@tearfund.org
Post (head office): 100 Church Road, Teddington TW11 8QE, UK

## Tyndall Centre for Climate Change Research

Brings together scientists, economists, engineers and social scientists who are working to develop sustainable responses to climate change through trans-disciplinary research and dialogue on both a national and international level.
www.tyndall.ac.uk
Tel: +44 (0) 1603 593900
Fax: +44 (0) 1603 593901
Email: tyndall@uea.ac.uk
Post (head office): Zuckerman Institute for Connective Environmental Research, School of Environmental Sciences, University of East Anglia, Norwich NR4 7TJ, UK

**World Bank**
One of the world's foremost financial institutions.
www.worldbank.org
The World Bank's climate change webpage is at:
http://go.worldbank.org/W13H8ZXSD1
Tel: (202) 473 1000
Fax: (202) 477 6391
Email: via the website
Post: The World Bank, 1818 H Street NW, Washington, DC 20433, USA

**Worldwatch Institute**
Provides information on the interactions among key environmental, social and economic trends revolving around the transition to an environmentally sustainable and socially just society.
www.worldwatch.org
Tel: 1 202 452 1999
Fax: 1 202 296 7365
Email: worldwatch@worldwatch.org
Post: 1776 Massachusetts Ave. NW, Washington, D.C. 20036-1904, USA

# Notes

## Introduction

1. Rose Tremain, *The Swimming Pool Season*, Vintage, 2003.
2. Muzelli's story is from Graham Erion, *Low Hanging Fruit Always Rots First: Observations from South Africa's Crony Carbon Market*, Center for Civil Society, University of KwaZulu-Natal, South Africa, 2005.
3. Joanna's story is from David Thomas, Henny Osbahr, Chasca Twyman, Neil Adger and Bruce Hewitson, *ADAPTIVE: Adaptations to Climate Change Amongst Natural Resource-dependant Societies in the Developing World: Across the Southern African Climate Gradient*, Tyndall Centre for Climate Change Research Technical Report No. 35, 2005.
4. *Reuters*, 17 August 2006.
5. Ibid.
6. Christian Aid, *The Climate of Poverty: Facts, Fears and Hope*, 2006, p. 12.
7. *Joint Science Academies' Global Response to Climate Change*, 7 June 2005, http://www.royalsoc.ac.uk/displaypagedoc.asp?id=20742.
8. Ibid.
9. Ibid.
10. Dr Paul Baer with Dr Michael Mastrandrea, *High Stakes: Designing Emissions Pathways to Reduce the Risk of Dangerous Climate Change*, Institute for Public Policy Research, 2006, p. 11.
11. *www.climatedenial.org*, 14 March 2007.
12. Jim Giles, 'Behind the Scenes', *Nature*, Vol. 445, 8 February 2007.
13. *Joint Science Academies' Global Response to Climate Change*.
14. J.T. Houghton, Y. Ding, D.J. Griggs, M. Noguer, P.J. van der Linden, X. Dai, K. Maskell, and C.A. Johnson (eds), *Climate Change 2001: The Scientific Basis. Contribution of Working Group I to the Third Assessment Report of the Intergovernmental Panel on Climate*

*Change*, Cambridge University Press, Cambridge, United Kingdom and New York, USA, p. 7.

15. World Meteorological Organization Press Release, 3 November 2006.

16. J.T. Houghton, Y. Ding, D.J. Griggs, M. Noguer, P.J. van der Linden, X. Dai, K. Maskell, and C.A. Johnson (eds), *Climate Change 2001: The Scientific Basis. Contribution of Working Group I to the Third Assessment Report of the Intergovernmental Panel on Climate Change*, Cambridge University Press, Cambridge, United Kingdom and New York, USA, p. 7.

17. *Guardian*, 11 October 2004.

18. NASA Goddard Institute for Space Studies, Research News, 25 September 2006.
    For examples of the 'hockey stick' curve, see http://www.ncdc.noaa.gov/paleo/pubs/millennium-camera.pdf. A slightly simplified version of the same graph is available on page 3 of http://www.grida.no/climate ipcc_tar/wg1/pdf/WG1_TAR-FRONT.PDF
    See also Michael E. Mann, Raymond S. Bradley and Malcolm K. Hughes, *Northern Hemisphere Temperatures During the Past Millennium: Inferences, Uncertainties, and Limitations*, February 1994, p. 12.

19. Rachel Warren, 'Impacts of Global Climate Change at Different Annual Mean Global Temperature Increases', in Hans Joachim Schellnhuber, Wolfgang Cramer, Nebojsa Nakicenovic, Tom Wigley and Gary Yohe (eds), *Avoiding Dangerous Climate Change*, New York, Cambridge University Press, 2006, p. 100.

20. Ibid, p. 100.

21. Ibid, p. 97.

22. *Associated Press*, 21 November 2006.

23. Ibid.

24. Ibid.

25. Simon L. Lewis, Oliver L. Phillips, Timothy R. Baker, Yadvinder Malhi and Jon Lloyd, 'Tropical Forests and Atmospheric Carbon Dioxide: Current Conditions and Future Scenarios', in Hans Joachim Schellnhuber, Wolfgang Cramer, Nebojsa Nakicenovic, Tom Wigley and Gary Yohe (eds), *Avoiding Dangerous Climate Change*, New York, Cambridge University Press, 2006, p. 151.

26. C. Turley, J.C. Blackford, S. Widdicombe, D. Lowe, P.D. Nightingale and A.P. Rees, 'Reviewing the Impact of Increased Atmospheric $CO_2$ on Oceanic pH and the Marine Ecosystem', in Hans Joachim

Schellnhuber, Wolfgang Cramer, Nebojsa Nakicenovic, Tom Wigley and Gary Yohe (eds), *Avoiding Dangerous Climate Change*, New York, Cambridge University Press, 2006, p. 65.

27. Ibid, p. 67.
28. Ibid, p. 67.
29. Ibid, p. 67.
30. *Reuters*, 21 September 2006.
31. *Environment News Service*, 17 February 2006.
32. James Hansen, 'Greenland Ice Cap Breaking Up at Twice the Rate It Was Five Years Ago, Says Scientist Bush Tried to Gag', *www.common dreams.org*, 17 February 2006.
33. *Eye on Earth*, 4 September 2006.
34. *Worldwatch Institute*, 6 November 2006.
35. Ibid.
36. *Associated Press*, 7 September 2006.
37. Ibid.
38. Ibid.
39. Fred Pearce, *The Last Generation: How Nature Will Take Her Revenge for Climate Change*, Eden Project Books, 2006, p. 109.
40. Ibid, p. 43.
41. James Hansen, 'Greenland Ice Cap Breaking Up at Twice the Rate It Was Five Years Ago, Says Scientist Bush Tried to Gag', *www.commondreams.org*, 17 February 2006,.
42. Andy Rowell, 'Climate Change Threatens Pacific Islands', *www.price ofoil.org*, 25 October 2006.
43. *Washington Post*, 29 January 2006.

## Chapter 1 Preparing to be prepared

1. W. Neil Adger, Nick Brooks, Graham Bentham, Maureen Agnew and Siri Eriksen, *New indicators of vulnerability and adaptive capacity*, Tyndall Centre for Climate Change Research Technical Report 7, 2004, p. 100.
2. Rajendra Pachauri, 'Avoiding Dangerous Climate Change: Presentation given to the Exeter Conference', in Hans Joachim Schellnhuber, Wolfgang Cramer, Nebojsa Nakicenovic, Tom Wigley and Gary Yohe (eds), *Avoiding Dangerous Climate Change*, New York, Cambridge University Press, 2006, p. 5.

3. *Joint Science Academies' Global Response to Climate Change*, 7 June 2005, http://www.royalsoc.ac.uk/displaypagedoc.asp?id=20742.

4. David Thomas, Henny Osbahr, Chasca Twyman, Neil Adger and Bruce Hewitson, *ADAPTIVE: Adaptations to climate change amongst natural resource-dependant societies in the developing world: across the Southern African climate gradient*, Tyndall Centre for Climate Change Research Technical Report 35, 2005, p. 5.

5. Lisa Schipper and Mark Pelling, *Disaster risk, climate change and international development: scope for, and challenges to, integration*, Blackwell Publishing, 2006, p. 28.

6. *Tiempo* Issue 49, September 2003.

7. Oxfam, *Adapting to climate change in Southern Africa*, http://www.oxfam.org.uk/what_we_do/issues/climate_change/story _climatechange.htm?searchterm=Adapting+to+climate+change+in+ Southern+Africa+.

8. Ibid.

9. David Thomas, Henny Osbahr, Chasca Twyman, Neil Adger and Bruce Hewitson, *ADAPTIVE: Adaptations to climate change amongst natural resource-dependant societies in the developing world: across the Southern African climate gradient*, Tyndall Centre for Climate Change Research Technical Report 35, 2005, p. 17.

10. Oxfam, *Adapting to climate change in Southern Africa* http://www.oxfam.org.uk/what_we_do/issues/climate_change/story _climatechange.htm?searchterm=Adapting+to+climate+change+in+ Southern+Africa+.

11. W Neil Adger, Nick Brooks, Graham Bentham, Maureen Agnew and Siri Eriksen, *New indicators of vulnerability and adaptive capacity*, Tyndall Centre for Climate Change Research Technical Report 7, 2004, p. 39.

12. Pablo Suarez, Chapter 2 – 'Predictions, Decisions and Vulnerability: Theoretical explorations and evidence from Zimbabwe', in *Decision-making for reducing vulnerability given new climate predictions: case studies from Metro Boston and rural Zimbabwe*, Boston University, April 2005, p. 35.

13. Saleemul Huq, *Adaptation to climate change: A paper for the International Climate Change Taskforce*, Institute for Public Policy Research, 2005, p. 8.

14. United Nations Development Programme Press Release, 25 July 2003.

15. *WORLDWATCH magazine*, March/April 2005, p. 19.

16. Rachel Warren, 'Impacts of Global Climate Change at Different Annual Mean Global Temperature Increases', in Hans Joachim Schellnhuber, Wolfgang Cramer, Nebojsa Nakicenovic, Tom Wigley and Gary Yohe (eds), *Avoiding Dangerous Climate Change*, New York, Cambridge University Press, 2006, p. 96.

17. Bill Hare, 'Relationship Between Increases in Global Mean Temperature and Impacts on Ecosystems, Food Production, Water and Socio-Economic Systems', in Hans Joachim Schellnhuber, Wolfgang Cramer, Nebojsa Nakicenovic, Tom Wigley and Gary Yohe (eds), *Avoiding Dangerous Climate Change*, New York, Cambridge University Press, 2006, p. 179.

18. *Worldwatch Institute*, 6 November 2006.

19. Bill Hare, 'Relationship Between Increases in Global Mean Temperature and Impacts on Ecosystems, Food Production, Water and Socio-Economic Systems', in Hans Joachim Schellnhuber, Wolfgang Cramer, Nebojsa Nakicenovic, Tom Wigley and Gary Yohe (eds), *Avoiding Dangerous Climate Change*, New York, Cambridge University Press, 2006, p. 179.

20. Ibid.

21. *Tiempo* Issue 54, January 2005, p. 19.

22. Ibid.

23. Ibid.

24. Andrew Simms, John Magrath and Hannah Reid with contributions from the Working Group on Climate Change and Development, *Up in smoke? Threats from, and responses to, the impact of global warming on human development*, New Economics Foundation, 2004, p. 29.

25. Ibid.

26. Ibid.

27. Ibid.

28. World Health Organisation, World Meteorological Organization and United Nations Environment Programme, *Climate Change and Human Health – Risks and Responses: Summary*, 2003, p. 5.

29. *Tiempo* Issue 57, October 2005, p. 28.

30. John Lanchbery, 'Climate Change-induced Ecosystem Loss and its Implications for Greenhouse Gas Concentration Stabilisation', in Hans Joachim Schellnhuber, Wolfgang Cramer, Nebojsa Nakicenovic, Tom Wigley and Gary Yohe (eds), *Avoiding Dangerous Climate Change*, New York, Cambridge University Press, 2006, p. 144.

31. United Nations, *Millennium Ecosystem Assessment*, 2005.
32. Ibid.
33. Ibid.
34. *Financial Times*, 22 March 2007.
35. Ibid.
36. Ibid.
37. John Magrath, *Glacier Melt: Why it Matters for Poor People*, Oxfam, 2004.
38. Rachel Warren, 'Impacts of Global Climate Change at Different Annual Mean Global Temperature Increases', in Hans Joachim Schellnhuber, Wolfgang Cramer, Nebojsa Nakicenovic, Tom Wigley and Gary Yohe (eds), *Avoiding Dangerous Climate Change*, New York, Cambridge University Press, 2006, p. 95.
39. Andrew Simms, Julian Oram and Petra Kjell, *The Price of Power: Poverty, Climate Change, the Coming Energy Crisis and the Renewable Revolution*, New Economics Foundation, 2004, p. 7.
40. Andrew Simms, John Magrath and Hannah Reid with contributions from the Working Group on Climate Change and Development, *Up in smoke? Threats from, and responses to, the impact of global warming on human development*, New Economics Foundation, 2004, p. 2.
41. C. Heslop-Thomas, W. Bailey, D. Amarakoon, A. Chen, S. Rawlins, D. Chadee, R. Crosbourne, A. Owino, K. Polsom, *Vulnerability to Dengue Fever in Jamaica*, Assessments of Impacts and Adaptations to Climate Change (AIACC) Working Paper No. 27, May 2006.
42. Pablo Suarez, abstract for the presentation *Reversed mainstreaming? Climate change and time preference in community disaster management* for the Development & Adaptation Days at the United Nations Conference of the Parties 11, Canada, 2005.
43. Ibid.
44. Excerpts of Decision 28/CP 7 from Annie Roncerel, Brook Boyer, Mozaharul Alam, '*Participatory Approaches for NAPA Preparation: An Overview*', p. 5.
45. Lobzang Dorji, abstract for presentation on *Experience with NAPA* for the Development & Adaptation Days at the United Nations Conference of the Parties 11, Canada, 2005.
46. Nicholas Stern, *The Economics of Climate Change: The Stern Review*, Cambridge University Press, 2007, p. 558.
47. Ibid, p. 112.

48. Correspondence with John Magrath of Oxfam, 10 April 2007.
49. Ibid.

## Chapter 2 False solutions

1. *Guardian*, 15 November 2006.
2. *Community Development Carbon Fund Annual Report 2004*, World Bank, p. 5.
3. *Environment News Service*, 11 December 2005.
4. Roberto C. Yap, *Can carbon reduction projects help reduce poverty? Two case studies from the Philippines*, presentation to the 'International Conference on the CDM – Linkages to Poverty Reduction and Sustainability', University of Copenhagen, October 2005.
5. *International Herald Tribune*, 20 December 2006.
6. Roberto C. Yap, *Can carbon reduction projects help reduce poverty? Two case studies from the Philippines*, presentation to the 'International Conference on the CDM – Linkages to Poverty Reduction and Sustainability', University of Copenhagen, October 2005.
7. *Tiempo* Issue 56, July 2005, p. 13.
8. World Bank Carbon Finance Unit, *The Role of the World Bank in Carbon Finance: An Approach for Further Engagement*, January 2006.
9. *Nature*, Vol. 445, February 2007.
10. Ibid.
11. Ibid.
12. *Down to Earth*, 8 March 2006.
13. Ibid.
14. Jessica Ayres, Maryanne Grieg-Gran, Lizzie Harris and Saleemul Huq, *Expanding the Development Benefits from Carbon Offsets*, the International Institute for Environment and Development, 2006.
15. World Bank, *State and Trends of the Carbon Market 2006*, May 2006, p. 18.
16. World Bank Press Release No:2007/146/SDN, 16 November 2006.
17. *ESI Africa* Issue 4/2003, 'Carbon Finance in Africa'.
18. Larry Lohmann, 'Marketing and Making Carbon Dumps: Commodification, Calculation and Counterfactuals in Climate Change Mitigation', *Science as Culture*, Vol. 14, September 2005, p. 217.
19. Ibid.

20. WWF, *The Gold Standard: Quality Assurance for CDM and JI Projects*, http://www.panda.org/about_wwf/what_we_do/climate_ change/solutions/ business_industry/offsetting/gold_standard/index. cfm, accessed 3 December 2006.

21. Graham Erion, *Low Hanging Fruit Always Rots First: Observations from South Africa's Crony Carbon Market*, Center for Civil Society, University of KwaZulu-Natal, South Africa, 2005, p. 42.

22. Ibid, p. 47.

23. Ibid, p. 41.

24. *Q&A with Mark Kenber (Senior Policy Officer for the WWF's Climate Change Programme),* Renewable Energy and Energy Efficiency Partnership, http://www.reeep.org/index.cfm?articleid=37&iid =904, accessed 3 December 2006.

25. Carbon Trade Watch, *Hoodwinked in the Hothouse: The G8, Climate Change and Free-market Environmentalism*, Transnational Institute briefing series No 2005/3, June 2005, pp. 46–7.

26. *BusinessWeek online*, 12 December 2005.

27. Larry Lohmann, 'Marketing and Making Carbon Dumps: Commodification, Calculation and Counterfactuals in Climate Change Mitigation', *Science as Culture*, Vol. 14, September 2005, pp. 216–17.

28. Ibid, p. 218.

29. *www.pointcarbon.com*, 19 February 2007.

30. Heidi Bachram, 'Climate Fraud and Carbon Colonialism: The New Trade in Greenhouse Gases', *Capitalism Nature Socialism*, December 2004, p. 6.

31. *www.pointcarbon.com*, 19 February 2007.

32. Ibid.

## Chapter 3 The World Bank

1. World Bank news, *Work on Investment Framework for Clean Energy and Sustainable Development Launched*, 24 September 2005.

2. Carbon Trade Watch, *Hoodwinked in the Hothouse: The G8, Climate Change and Free-market Environmentalism*, Transnational Institute briefing series No 2005/3, June 2005, p. 26.

3. Ibid.

4. Andrew Simms, Julian Oram and Petra Kjell, *The Price of Power:*

*Poverty, Climate Change, the Coming Energy Crisis and the Renewable Revolution*, New Economics Foundation, 2004, p. 27.

5. Ibid, p. 20.

6. Kate Hampton, *Catalysing Commitment on Climate Change*, Institute for Public Policy Research, 2005, p. 13.

7. *Striking a Better Balance: The Extractive Industries Review*, World Bank Group, December 2003.

8. 'Executive Summary' of *Striking a Better Balance: The Extractive Industries Review*, World Bank Group, p. 6.

9. Ibid, p. 10.

10. World Bank Group Management Response to *Striking a Better Balance – The World Bank Group and Extractive Industries: the Final Report of the Extractive Industries Review*, 17 September 2004, p. 7.

11. Statement by Peter Woicke, Managing Director, World Bank Group and Executive Vice President, International Finance Corporation at the opening session of the International Conference for Renewable Energies, Bonn, 2004.

12. Elizabeth Bast and David Waskow, *Power Failure: How the World Bank Is Failing to Adequately Finance Renewable Energy for Development*, Friends of the Earth – United States, October 2005, p. 12.

13. Ibid, p. 1.

14. Ibid, p. 1.

15. Ibid, p. 1.

16. Ibid, p. 14.

17. Katherine Sierra, Vice President for Infrastructure and Network Head, The World Bank, Keynote Speech to the International Grid-Connected Renewable Energy Policy Forum, Mexico, February 2006.

18. See *Media Briefing: Development Aid and the Baku–Ceyhan Pipeline*, Friends of the Earth, May 2005.

19. Greg Muttitt and James Marriott, *Some Common Concerns: Imagining BP's Azerbaijan–Georgia–Turkey Pipelines System*, Campagna per la Riforma della Banca Mondiale, CEE Bankwatch Network, The Corner House, Friends of the Earth International, The Kurdish Human Rights Project and PLATFORM, 2002, p. 159.

20. *Memorandum of the President of the International Development Association and the International Finance Corporation to the*

*Executive Directors on the Country Assistance Strategy for the People's Republic of Bangladesh*, World Bank Group, February 2001 quoted in Jim Vallette and Steve Kretzmann, *The Energy Tug of War: The Winners and Losers of World Bank Fossil Fuel Finance*, Sustainable Energy & Economy Network, April 2004, p. 10.

21. Jim Vallette and Steve Kretzmann, *The Energy Tug of War: The Winners and Losers of World Bank Fossil Fuel Finance*, Sustainable Energy & Economy Network, April 2004, p. 10.

22. Odin Knudsen, Senior Manager, Carbon Finance Unit, quoted in World Bank, *Carbon Finance Annual Report 2005*, p. 14.

23. World Bank, *Clean Energy and Development: Towards an Investment Framework*, April 2006, p. 115.

24. World Bank, *Environmentally and Socially Sustainable Development Network (ESSD) Reference Guide 1*, p. 11.

25. World Bank, *Environment Strategy Annex F*, 2001, p. 177.

26. CDM Watch, *The World Bank and the Carbon Market: Rhetoric and Reality*, April 2005, p. 7.

27. Ibid.

28. Ibid., p. 8.

29. Ibid.

30. *Financial Times*, 21 March 2007.

31. The World Bank Group and The Energy and Mining Sector Board, *World Bank Group Progress on Renewable Energy and Energy Efficiency Fiscal Year 2005*, World Bank, December 2005, p. xi.

32. International Rivers Network and Friends of the Earth International, Media Advisory, *World Bank 'New Investment Framework': A Great Leap Backwards for Sustainable Energy*, 6 December 2005.

33. *WORLDWATCH magazine*, May/June 2005.

34. Ibid.

35. Bretton Woods Project, *World Bank 'Pushing Big Dams' across Asia*, 20 January 2006.

36. *Environment News Service*, 19 April 2004.

37. Jon Sohn, Smita Nakhooda and Kevin Baumert, *Mainstreaming Climate Change Considerations at the Multilateral Development Banks*, World Resources Institute, July 2005, p. 8.

38. Ibid, p. 7.

39. World Bank, *Clean Energy and Development: Towards an Investment Framework*, April 2006, p. 120.

40. Ibid.

41. Correspondence with Antonio Hill.
42. E.L.F. Schipper, *Exploring Adaptation to Climate Change: A Development Perspective*, unpublished Ph.D. thesis, School of Development Studies, University of East Anglia; and E.L.F. Schipper *A Closer Look at Water Resources Management for Adaptation: Small-Scale Irrigation for Adapting to Droughts in Ethiopia*, draft paper presented at Development and Adaptation Days at COP-11, 4 December 2005.
43. Andrew Simms with support from Hannah Reid, *Africa: Up in Smoke? The Second Report from the Working Group on Climate Change and Development*, New Economics Foundation, June 2005, p. 2.
44. Saleemul Huq, *Adaptation to Climate Change: A Paper for the International Climate Change Taskforce*, Institute for Public Policy Research, 2005, p. 3.
45. Nicholas Stern, *The Economics of Climate Change: The Stern Review*, Cambridge University Press, 2007, p. 558.
46. Ibid.
47. Saleemul Huq, *Adaptation to Climate Change: A Paper for the International Climate Change Taskforce*, Institute for Public Policy Research, 2005, p. 8.
48. Nicholas Stern, *The Economics of Climate Change: The Stern Review*, Cambridge University Press, 2007, p. 558.
49. Ibid., p. 433.

## Chapter 4 Come together as one?

1. Quoted in Stephen H. Schneider and Janica Lane, 'An Overview of "Dangerous" Climate Change', in Hans Joachim Schellnhuber, Wolfgang Cramer, Nebojsa Nakicenovic, Tom Wigley and Gary Yohe (eds), *Avoiding Dangerous Climate Change*, New York, Cambridge University Press, 2006, p. 7.
2. 'Article 2: Objective', in *United Nations Framework Convention on Climate Change*, United Nations, 1992, p. 4.
3. International Energy Agency, *Deploying Climate-friendly Technologies through Collaboration with Developing Countries*, November 2005, p. 15.
4. Ibid.
5. Worldwatch Institute, *State of the World 2006*, Worldwatch Institute, 2006, p. 9.

6. *Financial Times*, 7 October 2005.

7. Tom Athanasiou and Paul Baer, 'The Science of Drawing the Line', *Climate Equity Observer*, http://ecoequity.org/ceo/ceo_6_2.htm.

8. Worldwatch Institute, *State of the World 2006*, Worldwatch Institute, 2006, p. 5.

9. *BBC News* website, 19 December 2005.

10. Christian Aid, *The Climate of Poverty: Facts, Fears and Hope*, 2006, p. 39.

11. Christian Aid, *Coming Clean: Revealing the UK's True Carbon Footprint*, 2007, p. 9.

12. For more on this see Ott, H., H. Winkler, B. Brouns, S. Kartha, M.J. Mace, S. Huq, A. Sari, J. Pan, Y. Sokona, P. Bhandari, A. Kassenberg, E. La Rovere and A. Rahman, *South-North Dialogue on Equity in the Greenhouse: A Proposal for an Adequate and Equitable Global Climate Agreement*, Eschborn, 2004.

13. *Associated Press*, 16 March 2007.

14. Ibid.

15. Ibid.

16. Ibid.

17. Ibid.

18. *Tiempo* Issue 58, January 2006, p. 25.

19. Ibid.

20. For more on the implications of technology see Section VII 'Technological Options', in Hans Joachim Schellnhuber, Wolfgang Cramer, Nebojsa Nakicenovic, Tom Wigley and Gary Yohe (eds), *Avoiding Dangerous Climate Change*, New York, Cambridge University Press, 2006, pp. 333–92.

21. Ibid., p. 333.

22. United Nations, *Convention on Biological Diversity*, 5 June 1992.

23. Rachel Warren, 'Impacts of Global Climate Change at Different Annual Mean Global Temperature Increases', in Hans Joachim Schellnhuber, Wolfgang Cramer, Nebojsa Nakicenovic, Tom Wigley and Gary Yohe (eds), *Avoiding Dangerous Climate Change*, New York, Cambridge University Press, 2006, p. 93.

24. Dr Paul Baer with Dr Michael Mastrandrea, *High Stakes: Designing Emissions Pathways to Reduce the Risk of Dangerous Climate Change*, Institute for Public Policy Research, November 2006, p. 4.

25. Ibid.

26. Ibid.

27. Ibid.
28. Ibid.
29. Andrew Simms, *Ecological Debt: The Health of the Planet and the Wealth of Nations*, Pluto Press, 2005, p. 159.
30. Ibid., p. 164.
31. Ibid., p. 156.
32. Ibid., p. 176.
33. Simon Roberts and Joshua Thumim, *A Rough Guide to Individual Carbon Trading: The ideas, the Issues and the Next Steps*, Centre for Sustainable Energy/Department for Environment, Food and Rural Affairs, November 2006, p. 3.
34. Ibid, p. 8.
35. *www.ecoequity.org*.
36. T. Athanasiou, S. Kartha and P. Baer, *Greenhouse Development Rights: An Approach to the Global Climate Regime that Takes Climate Protection Seriously While Also Preserving the Right to Human Development*, EcoEquity and Christian Aid, 2006, p. 3.
37. This explanation of Greenhouse Development Rights is paraphrased from Athanasiou, T., S. Kartha, P. Baer, *Greenhouse Development Rights: An Approach to the Global Climate Regime that Takes Climate Protection Seriously While Also Preserving the Right to Human Development*, EcoEquity and Christian Aid, 2006.
38. Ibid., p. 5.
39. Ibid., p. 4.
40. Gupta, Joyeeta, *'On Behalf of my Delegation, …' A Survival Guide for Developing Country Climate Negotiators*, Center for Sustainable Development of the Americas, 2000.
41. Jouni Paavola, 'Justice and Adaptation to Climate Change', *id21*, http://www.id21.org/insights/insights53/insights-iss53-art08.html.
42. Reuters, 11 May 2006.
43. CBS, 14 September 2006.

## Chapter 5 It's an issue of energy

1. Aaron Cosbey, Warren Bell and John Drexhage, *Which Way Forward? Issues in Developing an Effective Climate Regime after 2012*, International Institute for Sustainable Development, 2005, p. 8.
2. World Bank, *Clean Energy and Development: Towards an Investment Framework*, April 2006, p. 84.

3. Ibid.
4. US Department of Energy, *International Energy Outlook 2005*, July 2005, p. 8.
5. Ibid, p. 3.
6. Peter Schwartz and Doug Randall, *An Abrupt Climate Change Scenario and Its Implications for United States National Security*, Global Business Network, October 2003, p. 15.
7. *Global Public Media*, 29 July 2005.
8. Graham Strouts, 'What is Peak Oil?', on *www.transitionculture.org*, http://transitionculture.org/what-is-peak-oil/.
9. Ibid.
10. Ibid.
11. Robert L. Hirsch, Roger Bezdek and Robert Wendling, *Peaking of World Oil Production: Impacts, Mitigation, and Risk Management*, US Department of Energy, February 2005, p. 64.
12. Worldwatch Institute, *Renewables 2005 Global Status Report*, Renewable Energy Policy Network (REN21), 2005, p. 6.
13. Ibid, p. 19.
14. Sven Teske, Arthouros Zervos and Oliver Schäfer, *Energy [R]evolution: A Sustainable World Energy Outlook*, Greenpeace and the European Renewable Energy Council, January 2007, p. 4.
15. Reuters, 25 January 2007.
16. *Worldwatch Institute*, 11 January 2006.
17. World Bank, *Clean Energy and Development: Towards an Investment Framework*, April 2006, p. 50.
18. *Point Carbon*, 13 March 2007.
19. Ibid.
20 . *Financial Times*, 4 September 2005.
21. George Monbiot, *Heat: How to Stop the Planet Burning*, Allen Lane, 2006, p. 88.
22. *IPCC Special Report on Carbon Dioxide Capture and Storage*, Summary for Policymakers, September 2005, p. 18.
23. *Financial Times*, 22 January 2007.
24. *Financial Times*, 16 January 2007.
25. *Associated Press*, 16 November 2005.
26. *Reuters*, 4 March 2006.
27. *BBC News*, 1 September 2005.
28. *Environment News Service*, 20 February 2006.
29. Letters to the Editor on *www.grist.org*, 15 April 2005.

30. Worldwatch Institute, *State of the World 2006*, Worldwatch Institute, 2006, p. 11.
31. Helen Buckland, *The Oil for Ape Scandal: How Palm Oil Is Threatening Orang-Utan Survival*, Friends of the Earth, The Ape Alliance, The Borneo Orangutan Survival Foundation, The Orangutan Foundation (UK) and The Sumatran Orangutan Society, September 2005, p. 7.
32. Ibid, p. 13.
33. *Guardian*, 16 March 2006.
34. Ibid.
35. *Environment News Service*, 8 December 2005.
36. *Tiempo*, No. 51 April 2004, p. 23.
37. Helen Buckland, *The Oil for Ape Scandal: How Palm Oil Is Threatening Orang-Utan Survival*, Friends of the Earth, The Ape Alliance, The Borneo Orangutan Survival Foundation, The Orangutan Foundation (UK) and The Sumatran Orangutan Society, September 2005, p. 10.
38. Ibid.
39. *Reuters*, 23 August 2006.
40. *Reuters*, 7 March 2007.
41. Cowan Coventry, *Power to the People: Redirecting Energy Policy Towards the Poor*, speech to a Practical Action seminar, 17 July 2002.
42. Andrew Simms, Julian Oram and Petra Kjell, *The Price of Power: Poverty, Climate Change, the Coming Energy Crisis and the Renewable Revolution*, New Economics Foundation, 2004, p. 18.
43. Worldwatch Institute, *Energy for Development: The Potential Role of Renewable Energy in Meeting the Millennium Development Goals*, Renewable Energy Policy Network (REN21), 2005, p. 10.
44. Andrew Simms, Julian Oram and Petra Kjell, *The Price of Power: Poverty, Climate Change, the Coming Energy Crisis and the Renewable Revolution*, New Economics Foundation, 2004, p. 18.
45. United Nations Environment Programme, *Changing Climates: The Role of Renewable Energy in a Carbon-Constrained World*, Pre-Publication Draft, Renewable Energy Policy Network (REN21), December 2005, p. 27.
46. Andrew Simms, Julian Oram and Petra Kjell, *The Price of Power: Poverty, Climate Change, the Coming Energy Crisis and the Renewable Revolution*, New Economics Foundation, 2004, p. 29.

47. Andrew Simms, Julian Oram and Petra Kjell, *The Price of Power: Poverty, Climate Change, the Coming Energy Crisis and the Renewable Revolution*, New Economics Foundation, 2004, p. 4.

48. *SPIEGEL Magazine*, interview with China's Deputy Minister of the Environment, 7 March 2005.

49. Worldwatch Institute, *Feeling the Warming: Villagers in Southwestern China Grapple with Climate Change*, 19 December 2006.

50. Greenpeace News, 13 October 2005.

51. Reuters, 28 December 2006.

52. Sven Teske, Arthouros Zervos and Oliver Schäfer, *Energy [R]evolution: A Sustainable World Energy Outlook*, Greenpeace and the European Renewable Energy Council, January 2007, p. 82.

53. Ibid, p. 85.

54. Ibid, p. 22.

55. *Guardian*, 25 January 2006.

56. *Renewable Energy World*, 1 July 2004.

57. US Department of Energy Press Release, 29 July 2005.

58. Reuters, 28 February 2007.

59. Quoted from Rob Hopkins, *Energy Descent Pathways: Evaluating Potential Responses to Peak Oil*, MSc Dissertation for the University of Plymouth, 2006.

60. 'Why "Transition Culture"?', on *www.transitionculture.org*, http://transitionculture.org/why-transition-design/.

## Conclusion

1. Alice Walker, *Now is the Time to Open Your Heart*, Great Britain: Phoenix, 2005, p. 28.

2. *www.climatedenial.org*, 19 December 2006.

3. *Financial Times*, 4 January 2007.

4. *United Press International*, 7 December 2006.

5. *The Arizona Republic*, 7 December 2006.

6. Ibid.

7. *Financial Times*, 4 January 2007.

8. Peter Schwartz and Doug Randall, *An Abrupt Climate Change Scenario and Its Implications for United States National Security*, Global Business Network, October 2003, p. 4.

9. Ibid.

10. Ibid.

11. *RealClimate*, 8 January 2007.

12. Ibid.

13. Ibid.

14. Ibid.

15. Fred Pearce, *The Last Generation: How Nature Will Take Her Revenge for Climate Change*, Eden Project Books, 2006, p. 35.

16. Nicholas Stern, *The Economics of Climate Change: The Stern Review*, Cambridge University Press, 2007, p. 299.

17. See Part III of Nicholas Stern, *The Economics of Climate Change: The Stern Review*, Cambridge University Press, 2007.

18. Christian Aid News, 30 October 2006.

19. *Financial Times*, 10 November 2006.

20. Lester R. Brown, *Plan B 2.0: Rescuing a Planet Under Stress and a Civilization in Trouble*, Earth Policy Institute, 2006, p. 38.

21. Ibid.

22. Worldwatch Institute, *Vital Signs 2005*, Worldwatch Institute, 2005.

23. *Financial Times*, 2 November 2006.

24. Ibid.

25. Quoted in Carbon Trade Watch, *Hoodwinked in the Hothouse: The G8, Climate Change and Free-market Environmentalism*, Transnational Institute briefing series No 2005/3, June 2005, p. 11.

26. *Financial Times* 28 April 2006.

27. Bert Metz and Detlef van Vuuren, 'How, and at What Costs, Can Low-Level Stabilization be Achieved? An Overview', in Hans Joachim Schellnhuber, Wolfgang Cramer, Nebojsa Nakicenovic, Tom Wigley and Gary Yohe (eds), *Avoiding Dangerous Climate Change*, New York, Cambridge University Press, 2006, p. 344.

28. *Financial Times*, 10 March 2007.

29. *Dagens Nyheter*, 01 October 2005.

30. *Guardian*, 8 February 2006.

31. Paul Wapner and John Willoughby, 'The Irony of Environmentalism: The Ecological Futility but Political Necessity of Lifestyle Change', *Ethics & International Affairs*, Vol. 19, No. 3, November 2005.

32. Anderson K., Shackley S., Mander S. and Bows A., *Decarbonising the UK: Energy for a Climate Conscious Future*, Tyndall Centre for Climate Change Research Technical Report No. 33, 2005, p. 49.

33. Ibid.

34. George Marshall, *Denial and the Psychology of Climate Apathy*,

unpublished article. For more from George Marshall see *www.climate denial.org*.

35. Joanna R. Macy and Molly Young Brown, *Coming Back to Life: Practices to Reconnect Our Lives, Our World*, New Society Publishers, 1998, p. 37.

36. Ibid, p. 17.

37. Ibid.

38. Ibid, p. 86.

39. Ibid, p. 68.

40. Ibid.

# Index